# RESEARCH METHODOLOGY IN PLANT SCIENCES

# RESEARCH METHODOLOGY IN PLANT SCIENCES

**V.P.Singh**
**&**
**Shubhrata Purohit**

Institute of Environmental
Management and Plant Sciences,
Vikram University
Ujjain-456 010 (M.P)

SCIENTIFIC
PUBLISHERS

Published by
SCIENTIFIC PUBLISHERS (INDIA)
5 A, New Pali Road, P.O. Box 91
Jodhpur 342 001 (INDIA)
E-mail: info@scientificpub.com
Website: http://www.scientificpubonline.com

Print 2026



ISBN: 978-81-72333-15-7

Printed in India

# PREFACE

It is probably true to say that, though, many informations on the development of research analysis in plant sciences are scattered into the literature and number of methods are available for research analysis of plants, water and soils. But, the analyst is confronted by some problems. The first problem is, to select a procedure, which will properly evaluate the related parameter. The second problem is, to discover appropriate contents, which are present in the experimental samples (either plants or water or soil) one is attempting to analyse. The third problem is, to find the most specific methods, appropriate to the required analysis.

So, the aim of this book is, to acquaint the researchers through the collection of commonly used research methods for analysis of plants, water and soil samples. Specific methodology is given in the following portions in relation to a progressive analytical procedures :

I. Methodology for Plant Analysis.

II. Methodology for Water Analysis.

III. Methodology for Soil Analysis.

IV. Methodology for Ecological Studies and Vegetational Analysis.

The described methods have proved to be reliable analytical techniques.

**V.P. Singh**
**Shubhrata Purohit**

# CONTENTS

## PART – I
METHODOLOGY FOR PLANT ANALYSIS

## PART – II

### METHODOLOGY FOR WATER ANALYSIS

## PART - III
## METHODOLOGY FOR SOIL ANALYSIS



## PART – IV
## METHODOLOGY FOR ECOLOGICAL STUDY AND VEGETATIONAL ANALYSIS

## PART - V
## APPENDICES

# PART - I

# METHODOLOGY FOR PLANT ANALYSIS

**PART – I**

# METHODOLOGY FOR PLANT ANALYSIS

## 1. FRESH / DRY WEIGHT OF PLANTS

After growth of specific time duration, plants are collected and weigh these plants either whole or cut them into different plant parts like root, shoot, leaves and fruits etc. as a fresh matter. The weight of these fresh matter production indicate the fresh weight of plants.

After taking the fresh weight this matter is put in an oven at 80□C temperature for 72 hours. After that take weight again, this weight indicates dry weight of plants.

## 2. GROWTH PARAMETERS

Growth performance, such as, Relative growth rate (RGR), Net assimilation rate (NAR) and Leaf area ratio (LAR) can be determined as suggested by Blackman (1968). Different growth parameters are determined by the following formula —

*(i) Relative Growth Rate (RGR)*

RGR is the increase in dry weight per unit day weight per unit time.

$$\text{RGR (g/g/day)} = \frac{\text{Log } w_2 - \text{Log } w_1}{T_2 - T_1}$$

*(ii) Net Assimilation Rate (NAR)*

NAR is the increase in plant dry weight per unit leaf area in an unit time.

$$\text{NAR (g/cm}^2\text{/day)} = \frac{(w_2 - w_1)(\text{Log}_e A_2 - \text{Log}_e A_1)}{(T_2 - T_1)(A_2 - A_1)}$$

*(iii) Leaf Area Ratio (LAR)*

$$\text{LAR (cm}^2/\text{g)} = \frac{(A_2 - A_1)(\text{Log}_e\, w_2 - \text{Log}_e\, w_1)}{(\text{Log}_e\, A_2 - \text{Log}_e\, A_1)(w_2 - w_1)}$$

Where, $T_1$ and $T_2$ are the time of first and second harvesting.

$w_1$ and $w_2$ are total plant dry weight of time $T_1$ and $T_2$

$A_1$ and $A_2$ are the leaf area at time $T_1$ and $T_2$

## 3. LEAF AREA AND DRY WEIGHT RATIO

The area of the leaf is plotted and measured, after that the same leaves are kept at 80□C for 48 hours till a constant weight is obtained. The leaf area dry weight ratio is calculated as follows :

$$\text{Leaf Area Dry Weight Ratio} = \frac{\text{Total Leaf Area (cm}^2)}{\text{Dry Weight of leaf (g)}}$$

## 4. YIELD PARAMETER

On harvesting the crop, 3 or 5 plant replicates are taken for per plant yield attributes viz. number of pods or cobs per plant, number of seeds per pod or cob, number of seeds per plant seed weight and 1000 seed weight. Grain yield and harvest index are calculated on meter square area.

Harvest index is the ratio of grain yield per unit of biomass.

## 5. PTI (Percent Thermoinstability), PHSI (Plant Height Stress Index), DMSI (Dry Matter Stress Index), S (Drought Susceptibility Index), YS (Yield Stability Ratio)

These all parameters are calculated using the following formulae

**(1) Bausloma and Schapaugh, 1984**

$$\text{PTI} = 1 - \frac{1 - (T_1/T_2)}{1 - (C_1/C_2)} \times 100$$

Where, $T_1/T_2$ = Conductivity ratio under rainfed condition ($E_2$)

$C_1/C_2$ = Conductivity ratio under irrigated condition ($E_1$)

$$\text{PHSI} = \frac{\text{Plant height under } E_2}{\text{Plant height under } E_1} \times 100$$

$$\text{DMSI} = \frac{\text{Total dry matter produced under } E_2}{\text{Total dry matter produced under } E_1} \times 100$$

**(2) Fischer and Maurer, 1978**

$$S = 1 - \frac{Yd}{Yp} / D$$

Where,

Yd and Yp are the mean yield of the genotype under $E_2$ and $E_1$ conditions respectively and D is the drought intensity.

$$D = 1 - \frac{\text{Mean Yd of all the genotype under } E_2}{\text{Mean Yp of all the genotype under } E_1} \times 100$$

**(3) Lewis, 1954**

$$YS = \frac{\text{Grain yield under } E_2}{\text{Grain yield under } E_1} \times 100$$

## 6. MOISTURE CONTENT

The moisture content of the plant materials (leaf, seeds, flowers, etc.) is determined on fresh weight basis. Three replicates of plant materials are taken and fresh weight is determined. These materials are then kept for drying at 80°C in oven for 72 h. and then their dry weight is determined. The moisture content is determined as follows :—

$$(\%) \text{ Moisture} = \frac{\text{Plant fresh weight} - \text{Plant dry weight}}{\text{Plant fresh weight}} \times 100$$

## 7. RELATIVE WATER CONTENT

It is measured following Fletcher et al. (1988). Turgid weight is found by placing the cut ends of three 1 g batches of leaves in small beakers containing water. The beakers are enclosed in a humid atmosphere at room temperature (25°C) for 6 hours. The submerged ends of the leaves are blotted rapidly and thoroughly before weighing. The leaves are then transferred to an oven at 120°C for 48 hours and their dry weight is determined. The mean RWC value for the 3 replicates is calculated from the formula—

$$\text{RWC} = \frac{(\text{Fresh weight} - \text{dry weight})}{(\text{Turgid weight} - \text{dry weight})}$$

## 8. WATER POTENTIAL

*Measurement by Dew-Point Microvoltmeter*

Since the introduction of this Pressure-Volume (PV) analyzer has proven to be valuable tool in the evalutary plant water relation. It's main advantage is that a large number of important water relations parameter can be obtained by a single procedure. For example, the osmotic adjustment of plants to water deficit or salt stress can easily be reduced by comparing the osmotic potential at full turgor and at zero turgor during a drought cycle or after subjecting the plants to salt stress. Furthermore, the bulk modulus of elasticity and the water storage capacity of the tissue investigated can be calculated and their dependency on changes in relative leaf water deficit modelled with a pressure-volume analysis a Hofter diagram can be constructed.

***Apparatus*** : L-51 Psychrometer/hygrometer (Wescor Le Logan USA).

*Modification in Instrument* : Instrument is slightly altered as recommended by Shackel (1984).

The commercially available aluminium block was well insulated with plastic form and covered with aluminium foil to improve refleehan. Additionally, about 5 cm of the electrical lead is insulted by covering it with the same form in a PVC tube. To simply weighing during the PV measurement. The wire connecting the in situ hygrometer to the micro volt meter is cut about ten cm from the sensor and fitted with the connector to allow the reattachment to the receiver of the wire. Dew-Point hygrometer is calibrated with NaCl solution of known concentration.

***Method*** : At least two hours before the experiment is started, for the *in situ* observation, hygrometer is placed into growth chamber to ensure temperature equilibration. According to the HR-33T instruelean mameal, the cooling coefficient ^v is at for drug, sealed hygrometer. Prior to the final installation the leaf is inserted into the leaf clamp. After making the shape of the aluminium specimen holder with small strips of taps, the leaf is removed. If the position of the hygrometer sensor the cutiles is gently separated away with emery paper (No. 600) to increase the conductance for water vapour exchange. The leaf is singed with distilled water and carefully blotted dry. A thin band of a mixture of Lavolin/beeswax (50 : 50) is then applied around the seeling face of the hygrometer sensor. The sensor is then attached to the leaf in the clamp by gently pressing it against the leaf surface. Inspite of abrading the cuticle, initial water vapour equilibration generally required up to three hours as determined by Psychrometer measurement. The Dew-Point hygrometer remained attached for several days without causing obvious changes to the leaf.

The Dew-Point hygrometric determination of water potential is interrupted by a Psychrometric measurement. The leaf is carefully cut off from the twig, the lead is disconnected from the microvolt meter and the leaf is weighed on an electrical balance. It is then put back into the chamber, reconnected to the microvolt meter and another hygrometer measurement is made. After an additional Psychrometric determination, while the leaf is allowed to transpire, the resulting changes in the leaf water potential would be monitored. This measuring cycle is repeated upto 20 times, until leaf water potential is for beyond the turgor loss point, which lasted more than 7 hours.

Finally the leaf is removed from the holder, cleaned and leaf area is determined by leaf area meter and keep in ovan for dry matter determination.

## Data Sheet

*Dew-Point hygrometer readings*

Date :

Name of Species :

Initial weight of leaf :

Weight of leaf after experiment :

Weight of whole arrangement + Leaf :

Weight of arrangement of leaf :

Leaf area :

**TABLE**

Sensor No.

| SN | PSR | Wt Whole Arr | Time | Reg. Analysis |
|---|---|---|---|---|
| | | | | |

Regression Analysis of Sensor (1)

$$L\text{-}51\text{-}I \text{ m Osmal} = -39.7951 + 53.1371 \times \mu V$$

## 9. OSMOTIC POTENTIAL

*(A) Determination of Osmotic Potential of cell-sap by Scholander-Pressure Pump*

***Method*** : Leaves of similar age are collected in aluminium foil and bring to them in laboratory and initial fresh weight is determined and then the leaves are kept in Scholander pressure vessel. The pressure chamber comprises an aluminium, stele or stainless steel pressure vessel that can withstand pressure upto 60 MPa connected to a pressurized supply of inert gas (usually nitrogen or compressed air) through a pressure reducing valve and metaring value and connected to pressure gas.

The top of the chamber can be reported from the cap [K] so that chamber tops quickly interchanged. Different tops may be necessary for grass leaves, leaves with large petioles, leaves with very short petioles or when several samples are measured. The seal [L] itself is important. It is made up of rubber that is sufficient elastic to fill in the indenlatians of irregularly shaped petioles, but not so soft that is disintegrates under pressure. For very irregular petioles, quick setting silicon compound can be used, but this shows down the numbers of leaves that can be measured.

***Theory*** : *Scholander (1964, 1965)* described the principles of the pressure chamber techniques. They showed that due to the evaporation of water from leaf cells by transpiration, coupled with resistances to flow of water from soil to the leaf, a negative hydrostatic pressure build up in the xylem. When a transpiring leaf is cut, the xylem sap recades in the xylem until it becomes restricted by a cross wall. Scholander et al. (1964) suggested that pressurizing the leaf until water just returns to the cut surface of the xylem tissue gives a measure of the hydrostatic pressure in the xylem.

Scholander et al. (1965) showed that once the cells in the tissue reach zero turgor pressure, Pe is related to the cell water —

$$1/Pe = 1 \# S = V_S - V/RT_n$$

Where, S = is the volume of symplastic water in the turgid leaf.

#S = is the osmotic pressure of the symplast

V = is the volume of the water lost from the leaf

R = is the Universal gas constant

T = is the absolute temperature and

n = is the number of solute in the symplast

**Study of the Pressure - Volumes relations**

Another advantage of the pressure chamber techniques is that it can be used to measure the pressurevolume relations to plant tissue. Scholander showed that if additional pressure [Pe] is applied to a leaf or twig beyond the initial balancing pressure, xylem sap can be collected and the volume determined. By doing this in stepwise manner, a pressurevolume curve can be established. Once the turgar pressure reached zero, the plot of 1/Pe against V becomes linearas predicted eq. 1. Establishing the pressurevolume curves of species has become popular becomes a wide range tissue water accumulate can be known.

*(B) Determination of Osmotic Potential of cell-sap by VAPRO —VAPOUR PRESSURE OSMOMETER*

The VAPRO Osmometer is an advanced electronic adaptation of the hygrometric method of vapour pressure determination. In this instrument, the sensitive thermocouple and sophisticated electronics provide the means to measure the dew-point temperature depression of a specimen with resolution to 0.00031□C.

Vapour pressure and freezing point are among the colligative properties of a solution. Compared with pure solvent, these properties are altered in proportion to the number of solute particles dissolved in each kilogram of solvent (water in the case of biological solutions). Thus, measuring either property is an indirect means of determining solution concentration or osmolality.

The chief advantage of the vapour pressure method is that it does not require alteration of the physical state of the specimen.

*Concomitant benefits include :*

- 10 microliter sample size.
- Routine operation on microsamples of any biological solution, including whole blood, serum, plasma, urine and sweat, as well as complex specimens such as tissue samples.
- None of the measurement artifacts that arise in freezing point depression measurements due to elevated viscosity, particulate matter, inhomogeneities, or other physical characteristics of the sample.
- Superior reliability because the measurement involves minimal mechanical complexity.

***Theory of Operation :*** Osmolality is an expression of the total concentration of dissolved particles in a solution without regard for particle size, density, configuration, or electrical charge. Indirect means for the measurements of osmolality are afforded by the fact that the addition of solute particles to a solvent changes the free energy of the solvent, i.e., vapor pressure, freezing point, and boiling point. Compared with pure solvent, the vapor pressure and freezing point of a solution are lowered, while its boiling point is elevated, provided that a single solvent is present in the solution. Solutions containing more than one solvent generally behave in more complex ways.

In single-solvent solution, the relative changes in solution properties are linearly related to the number of particles added to the solvent, although not necessarily linearly related to the weight of solute, since solute molecules may dissociate into two or more ionic components. Since these properties all change linearly in proportion to the concentration of solute particles, they are known as "colligative" properties.

Osmotic pressure is also a colligative property of a solution, but unlike the other three, it is not a cardinal property of the solvent. Solution osmotic pressure can be measured directly using a semipermeable membrane apparatus, but only with respect to those solute particles that are impermeable, since smaller solute particles freely transude the membrane and do not directly contribute to osmotic pressure. Such a measurement is referred to as "colloid osmotic pressure" or "oncotic pressure". It is expressed in terms of pressure, in mmHg or kPa. Total osmotic pressure, i.e., that which can be calculated on the basis of total solute concentration, is a theoretical concept only.

The measurement of total solution concentration, or osmolality, can only be made indirectly by comparing one of the solution colligative properties with the corresponding cardinal property of the pure solvent. The first practical laboratory instruments developed for routine measurement of osmolality were based upon depression of the freezing point and, until recent years, all osmometers for large-scale testing were based upon this methodology.

The VAPRO Osmometer embodies newer technology. It is based upon a measurement of vapour pressure depression made possible by thermocouple hygrometry. The vapour pressure method enjoys a significant intrinsic advantage over the measurement of either freezing point depression or boiling point elevation in that it it can be performed without the necessity for a change in the physical state of the specimen. It is thus a passive technique of measurement that is free from measurement artifacts that often occur when the specimen to be tested must be altered physically. This fundamental difference in methodology gives rise to the many advantages of the vapor pressure osmometer over the older method.

In the VAPRO Vapour pressure osmometer, a 10 microliter sample of the solution to be tested is pipetted onto a small, solute free paper disc which is then inserted into a sample chamber and sealed. A thermocouple hygrometer is incorporated integrally within the chamber. This sensitive temperature sensor operates on the basis of a unique thermal energy balancing principle to measure the dew point temperature depression within the chamber. This parameter, in itself a colligative property of the solution, is an explicit function of solution vapor pressure.

*(1) Equilibration and zero set* : The sample is introduced into the chamber and the chamber is closed. Simultaneously, "In Process" and a countdown by seconds is displayed. At this point, there will generally be some difference between the temperature of the specimen and the temperature of the sample chamber. Temperature equilibrium occurs within a few seconds. The vapor pressure may also reach equilibrium during this interval. The microvoltmeter reads the amplifier voltage to establish the reference for the measurement.

*(2) Cooling* : An electrical current is passed through the thermocouple, cooling it by means of the Peltier Effect to a temperature below the dew point. Water condenses from the air in the chamber to form microscopic droplets upon the surface of the thermocouple.

*(3) Dew-point convergence* : Electronic circuitry "pumps" thermal energy from the thermocouple via Peltier cooling in such a way as to cancel out heat influx to the thermocouple by conduction, convection, and radiation. Given this, the temperature of the thermocouple is controlled exclusively by the water condensing upon its surface. Thermocouple temperature depressed below the dew point in step @, rises asymptotically toward the dew point as water continues to condense. When the temperature of the thermocouple reaches the dew point, condensation ceases, causing the thermocouple temperature to stabilize.

*(4) End of sequence and readout* : The reading on the display is proportional to the vapour pressure of the solution. When this final reading is reached, a chime sounds and the "In Process"

changes to "Osmolality". The result is displayed in SI units of osmolality-mmol/kg.

## 10. MEMBRANE PERMEABILITY

It is measured following Fletcher and Drexlure (1980). The per cent leakiness of the leaves are determined by placing 3 leaf discs (1 $cm^2$) in 15 ml of distilled water for 24 hours at 24°C. The conductivity of the water is determined using conductance meter. The samples are then placed in boiling water-bath for 20 minutes, cool to 24°C and conductivity is determined again. The membrane permeability is calculated as per cent leakiness as the ratio of conductivity after 24 hours to conductivity after boiling.

## 11. MEMBRANE PHOSPHOLIPID

***Solutions :***

Conc. Perchloric acid : Store in acid bottle away from organic solvents.

| | | |
|---|---|---|
| ANS solution | : | 30 g Sodium Bisulfite $NaHSO_3$<br>6 g Sodium Sulfite $Na_2SO_3$<br>5 g 1,2,4-Aminopthal Sulfonic Acid (ANS)<br>250 ml dd water<br>Keep refrigerated in dark bottle. Will keep no more than 3 months. |
| 5% Ammonium Molybdenate | : | 5 g Ammonium Molybdenate $(NH_4)_6$ Mo7024 100 ml dd water Keep refrigerated. |
| 1mM Phosphate Std. | : | 0.012 g Sodium Phosphate Monobasic $NaH_2PO_4$ 100 ml dd water |
| 2:1 Chloroform :Methanol | : | 200 ml Chloroform, 100 ml Methanol, Use in fume hood. |
| 7% Saline Solution | : | 7 g Sodium Chloride NaCl, 100 ml dd water. |

***Extract Preparation*** : Grind plant material in 2:1 Chloroform : Methanol. Centrifuge debris at 3000 rpm for 10 minutes. Remove supernatant and place in corex tube. Add 1 ml 7% NaCl (aqueous) for every 5 ml supernatant and vortex. Centrifused at 10,000 rpm for 10 minutes. Remove and store lower supernatant.

***Phospholipid Oxidation*** : Place organic extract containing membranes in graduated 10 ml test tube. Dry under nitrogen to complete dryness. No solvent must be left. Add 0.5 ml of Conc. Perchloric acid and top the tube with a marble. In fume hood, heat in dry block at high temperature for 15 minutes or until solution is clear. Remove from heat and allow to cool for 10 minutes before removing the marbel. Do not touch with naked hand.

***Phospholipid Analysis*:** With the samples, prepare 2 more tubes one marked blank and the other marked standard. Add 0.5 ml Conc. perchloric acid and 1.0 ml of phosphate solution to the standard tube. Add 1 ml of dd water to all of the tubes. Add 0.4 ml ANS solution and 0.5 ml ammonium molybdate to all tubes. Vortex all tubes and incubate at room temperature for 15 minutes. Top all tubes to 10 ml with dd water and vortex. Determine the absorbance at 660 nm blanking on the blank tube. Determine phospholipid concentrations by ratio comparision with the standard :

$$[\text{Phospholipid}]\ \mu M = \frac{Apl}{Astd}$$

## 12. POROMETRIC STUDY

The measurement of stomatal behaviour is fundamental to most areas of plant research. However, these measurements have not always been particularly easy to get, usually involving lessthan-less convenient devices and procedures. Porometer is the instrument which is used to measure the stomatal resistance, transpiration rate, relative humidity, leaf temperature and the quantum received by the intact leaves, instantly. Porometers in general use today fall into two broad categories, transient or steady state, according to the measurement principle on which they are based. Transient porometers may be ventilate, in which a fan provides rapid chamber mixing, or unventilated, in which cuvette humidity changes occur strictly by diffusion. In ventilated transient porometers, stomatal resistance is calculated from the rate and magnitude of change in cuvette humidity, and system volume. In unventilated transient porometers, resistance is not computed directly from the magnitude of changes in relative humidity. Four types of ventilated porometers will be considered: these include the flow-through steady state porometer, automatic null balance steady state porometer, and ventilated transient porometer; the fourth type, a recirculating manually controlled null-balance porometer has some of the characteristics of each of the first three types.

***Theory/Principle :*** Water loss from a leaf placed in the porometer cuvette is determined by measuring the flow rate of dry air necessary to maintain a constant relative humidity inside the cuvette. Typically, the ambient relative humidity is used as a null point, and dry air is injected into the cuvette at a rate which is just sufficient to balance the transpirational water flux out of the leaf. This maintains the cuvette relative humidity at the set-point. Stomatal resistance (or conductance) is calculated directly from the measured values of relative humidity, leaf and air temperature, and flow rate.

***Procedure :*** The aperture area depending on leaf surface (broad/thin) and atmospheric pressure is initially entered into the porometer. After the cuvette is acclimated to ambient conditions, cuvette relative humidity is stored in memory with the help of HUM SET switch. the intact leaf sample is clamped in sensor head. The null adjust valve is used to bring the flow rate within the dynamic range of the automatic flow controller. After thirty seconds, hold switch is used to hold the various parameters in memory for display.

## 13. CHLOROPHYLL CONTENT

100 mg of fresh leaves are homogenised in 80 % acetone and centrifuged for 10 minutes at 3500 rpm. Then the supernatant is made upto 10 ml and the absorbance is read as respected following methods. These are some important methods which are commonly used in research work.

### (i) Arnon [1949]

Chlorophyll a (mg/g FW) = 12.7 OD 663 - 2.69 OD 645

Chlorophyll b (mg/g FW) = 22.9 OD 645 - 4.68 OD 663

The values for Chl a and Chl b are added together to obtain the quantity for total chlorophyll.

### (ii) Strain and Svec [1966]

Chl a ($\mu g\ ml^{-1}$) = 11.63 ($A_{665}$) - 2.39 ($A_{649}$)

Chl b ($\mu g\ ml^{-1}$) = 20.11 ($A_{649}$) - 5.18 ($A_{665}$)

Total Chl ($\mu g\ ml_{-1}$) = 6.45 ($A_{665}$) + 17.72($A_{649}$)

## 14. CAROTENOID CONTENT

For carotenoid estimation 100 mg of fresh leaf tissue is taken and crush in a pestle and mortar with 80% (v/v) Acetone and a pinch of clean, fine sand. The extract is centrifuged at 2500 g for 10 minutes and the clear supernatant is made upto 10 ml with 80% Acetone. The absorbance of the extract is read at 440 nm.

*Total carotenoids are estimated following the method given by Ikan (1969).*

$$\text{Total carotenoid (mg g}^{-1}\text{ F.W.)} = \frac{A}{196 \times L \times W}$$

Where, A = absorbance at 440 nm

L = diameter of spectrometer quartz cell, and

W = g sample / total volume in ml.

When the absorbance of extract is read at 663, 645 and 480 nms, the carotenoid content can be determined using Kirk and Allen (1965) following the formula —

$$Car_{480} = \Delta A_{480} + 0.114\ \Delta\ A_{663} - 0.638\ \Delta\ A_{645}$$

## 15. ANTHOCYANIN LEVEL

Anthocyanin comprise of a group of glycosides pigments responsible for various colours, particularly red, violet and blue in flower, stem, leaves and roots (beet root) of plant. They are soluble in water and generally occur in the cell-sap.

Anthocyanin is measured using whole plant materials by the method of Arummn and Mohr (1978). Plant material is extracted in 5 ml of HCl : Methanol (1 : 99 v/v). The extraction is carried out in darkness at low temperature. The absorbance is measured at 480, 510 and 525 nm in double beam spectrophotometer. Anthocyanin is measured as OD/g material.

## 16. HILL ACTIVITY

*This can be measured by using DCPIP Reduction Method.*

***Reagent required***

*(A) Composition of suspension medium*

(1) 1M Sorbitol 10 ml

(2) 0.1M $K_2HPO_4$ 7.5 ml

(3) 25mM NaEDTA 6.0 ml

(4) 10mM $NaHCO_3$ 6.0 ml

(5) Distilled water 0.5 ml

(pH should be adjusted at 7.6)

*(B) Composition of Assay medium*

(1) 1M Sorbitol 15 ml

(2) 75mM Tricine 12 ml

(3) 25mM NaEDTA 2.4 ml

(4) 10mM $NaHCO_3$ 12 ml

(5) Distilled water 18.6 ml

(pH should be adjusted at 7.6)

***Isolation of Chloroplast*** : Take one gram leaf tissues without major veins and grind them in a mortor with a little suspension medium. Make-up the extract to about 30 ml. Filter the extract through four layers of cheese/muslin cloth lined in a funnel. Make up the extract to 30 ml using the suspension medium. Centrifuge at 10,000 rpm at 0□C for 5 minutes. Decant the supernatant and resuspend the chloroplast by adding suspension medium and make up to 6 ml volume and keep it in a freeze. This is an isolated chloroplast suspension.

***Assay of Hill-reaction*** : For estimating the Hill-reaction, dilute the chloroplast suspension using the extraction medium (medium A) so as to get a concentration of 150 μg/ml chlorophyll. Take 2.8 ml of assay medium (medium B) in a cuvvette, add 0.1 ml of diluted chloroplast suspension and 0.1 ml of 1mM DCPIP, mixed well by inverting the cuvvette and the absorbance is measured immediately at 600 nm. Next, expose the cuvvette to the light source (using water filter between light source and cuvvette) immediately thereafter for a 5 minutes and measure the absorbance again. Now, by plotting the absorbance values against the standard graph for DCPIP, the amount of DCPIP in the reaction mixture before and after the illumination are be calculated.

The Hill-Reaction is estimated following the further formula—

μmol DCPIP reduced = μmol DCPIP in reaction mixture – μmol DCPIP remaining after illumination

$$\text{Hill-Activity} = \frac{\mu\text{mol DCPIP reduced} \times 60}{\text{illumination time in minutes} \times \mu\text{g chl in the reaction mixture}}$$

## 17. PROTEIN CONTENT

100 mg leaf tissue is crushed in 0.05 N NaOH solution with a pinch of clean fine sand. The homogenate is centrifuged at 3000 rpm for 10 minutes and the supernatant is made upto 10 ml using 0.05 N NaOH.

*The protein content in the extract is estimated following Lowry et el. (1951) as modified by McDonald and Chen (1965).*

### *Reagents*

1. 2% (w/v) $Na_2CO_3$ dissolve in 0.1 N NaOH
2. 2.7% (w/v) sodium potassium tartrate
3. 1% $CuSO_4$
4. 1N Folin-ciocalteu's phenol reagent dilute (1:1) with distilled water from a 2N commercial preperation, and make fresh always.

   0.6 ml of the extract is taken in a centrifuge tube and dilute with 2.4 ml of distilled water. To this is added 4.5 ml of 10% Trichloroacetic acid (TCA) to participate the protein fraction of the extract. The mixture is centrifuged, the supernatant discards, and the pellet is dissolved in 0.5 ml of 1N NaOH and allow to stand for 2 hours for the protein to dissolve. To this is added 5 ml carbonate-copper solution (prepare by adding 0.05 ml 2.7% sodium potassium tartrate and 0.05 ml 1% $CuSO_4$ into 4.9 ml of 2% $Na_2CO_3$, per test tube. This mixture is prepared in a quantity sufficient for all tubes and while preparing, the $CuSO_4$ solution is always added last. The mixture is always made freshly and incubated for exactly 10 minutes. At the end of 10 minutes 0.5 ml of Folin-phenol reagent is added, mix rapidly and allow to stand for 30 minutes at room temperature. The absorbance of the clear blue solution is noted at 700 nm and the protein content estimate from a standard curve prepare using Bovin Serum Albumin (BSA).

## 18. FREE PROLINE

*Free proline is estimated by a simple spectrophotometer method given by Bates et al. (1973).*

***Reagents***

1. 3% (w/v) sulfosalicylic Acid : Sulfosalicylic acid (3%) is prepared by 3 g of sulfosalicylic acid in 97 ml of distilled water.
2. Acid Ninhydrin : The reagent is prepared by mixing 20 ml of 6M orthophosphoric acid with 30 ml of glacial acetic acid. This mixture is heated at 60□C and 1.25 g Ninhydrin is dissolved slowly in this mixture.
3. Toluene.

***Procedure*** **:** 200 mg fresh leaf tissue is extracted for proline in accordance with Bates et al. (1973) by homogenizing in 10 ml of 3% aqueous sulfosalicylic acid and the homogenate is centrifuged at 5000 rpm for 10 minutes. Two ml of supernatant is reacted with 2 ml of acid ninhydrin. The tubes are incubated at 100□C for 1 hour in waterbath and later transferred into an icebath to terminate the reaction. Mixture extracted with 4 ml of toluene and brought to room temperature. The tubes are agitated vigorously for 15-20 seconds. The absorbance of this Coloured extract is measured at 520 nm using UV Visible Spectrophotometer against a Toluene blank.

*The amount of proline in the sample are determined using a standard curve against L-Proline.*

## 19. TOTAL FREE AMINO ACID

*For the extraction and estimation of total free amino acids, the method of Noorudeen and Kulandaivelu (1982) is followed.*

*(a) Extraction of aminoacids from the leaf tissue*

Preweighed material (200 mg) is extracted twice with 80% acetone (total volume 5 ml) after vigorous grinding using a pestle and mortar. The supernatant is used for the estimation of amino acids.

*(b) Estimation of amino acids*

Preparation of potassium cyanide-pyridine reagent : 2 ml of 0.01% KCN (10 mg of KCN dissolve in 100 ml of distilled water) and 1 ml of 80% phenol (4g dissolve in 5 ml of distilled water) is added to 100 ml pyridine.

0.5 ml of extract is mixed with 1 ml of KCN-pyridine reagent after which it is kept in a boiling water bath for 10 minutes. To this mixture 0.2 ml of 0.5% ninhydrin in absolute alcohol is added and the tubes are kept in boiling water bath for another 15 minutes. The tubes are cooled to room temperature, the volume is adjusted to 10 ml with 60% ethanol and the absorbance read at 575 nm. Triosene is used as a standard for the estimation of amino acids and is calculated as follows :

$$C = \frac{K \times A + B \times \text{dilution}}{\text{Fresh weight of the tissue (g)}}$$

Where, C = Concentration of total free amino acids in the sample

K = 115.19

B = 1.36

A = Absorbance of the sample at 575 nm

Amino acid content is expressed in mg/g of fresh weight.

## 20. STARCH CONTENT

*(a) Extraction of starch from leaf tissue*

Preweighed material (200 mg) is homogenised with 10 ml of 80% alcohol in a pestle mortar and centrifuged (3500 g) for 15 minutes. The supernatant is discarded and the pelette is again washed with 10 ml of 80% alcohol and recentrifuged. The supernatant is again discarded and the pellet is dissolved in 10 ml of distilled water and boiled for 40 minute in a water bath to dissolve the starch. The volume is made-up-to 10 ml and the supernatant is collected for starch estimation.

*(b) Estimation of starch*

Starch is estimated by the method given by Plummer (1971) with the help of iodine-potassium-iodide ($I_2KI$) reagent.

*(c) Preparation of $I_2KI$ reagent*

60 mg of Iodine dissolve in 10 ml of 6% potassium iodide solution. From this 1 ml is taken and added in distilled water along with 0.43 ml of hydrochloric acid (0.05 N) and the total volume is made upto 100 ml with distilled water.

The reaction mixture consist of—

1 ml - Starch extract

4 ml - Distilled water

1 ml - $I_2KI$ reagent

Absorbance of the reaction mixture is measured at 640 nm. The starch content is calculated from calibration curve prepare by using starch as a standard and it is calculated as follows —

$$C = \frac{K \times A + B \times \text{dilution}}{\text{Fresh weight of the tissue (g)}}$$

Where, C = Concentration of starch in the sample

K = 4014.7

B = – 0.1

A = Absorbance of the sample at 640 nm

Starch content is expressed in mg/g of fresh weight.

## 21. TOTAL SUGAR AND REDUCING SUGAR

*(a) Extraction of total sugar and reducing sugar from the leaf tissue*

Total free sugar (reducing plus non-reducing) is estimated by the method of Fong et al., 1953 (cf : Hassid and Abraham, 1957). Preweighed (200 mg) material is boiled for 10 minutes in 10 ml of 80% (v/v) chilled ethyl alcohol and the supernatant after centrifugation (3250 g, for 15 minutes) is used. The residue is again ground with 10 ml of 80 % (v/v) ethyl alcohol and centrifuge at 3250 g for 15 minutes. The two supernatants are combined and the volume make up to 20 ml. From this alcohol extract, 5 ml is taken for acid hydrolysis with 1N HCl (5 ml) in a boiling water bath for 40 minutes. This hydrolysed extract is used for the estimation of total sugar. From the above alcohol extract, 5 ml is taken and evaporate to dryness, the dry contents are dissolved in distilled water (10 ml)and cool in refrigerator for 30 minutes, this extract is used for the estimation of reducing sugars.

*(b) Estimation of total sugar and reducing sugar*

Preparation of Anthrone reagent : 200 mg of anthrone is dissolved in 100 ml of chilled concentrated sulphuric acid.

The reaction mixture consist of :

2 ml - Total sugar or reducing sugar extract

4 ml - Anthrone reagent

The reaction mixture is kept in boiling water bath for 7.5 minutes to develop colour, the absorbance is taken at 640 nm after cooling.

Known concentration of glucose is used as standards and with the help of glucose calibration, total sugar and reducing sugar are calculated as follows :

$$C = \frac{K \times A + B \times \text{dilution}}{\text{Fresh weight of the tissue (g)}}$$

Where : C = concentrationof total or reducing sugar in the sample

K = 111.70

B = – 2.12

A = Absorbance of the sample at 640 nm.

Sugar content is expressed in mg/g of fresh weight.

*(c) Estimation of total sugar (Malhotra and Sarkar, 1979)*

***Reagents***

(i) Copper reagent A : Dissolve 25 g of sodium carbonate (anhydrous), 25 g of Rochelles salt (SPT), 20 g of sodium bicarbonate and 200 g of sodium sulphate in 800 ml of distilled water and dilute it to one litre. Filter and store.

(ii) Copper reagent B : 15% of $CuSO_4.7H_2O$ containing 1 or 2 drops of conc. $H_2SO_4$ per 100 ml.

(iii) 1N $H_2SO_4$

(iv) 1N NaOH

(v) Arseno Molybdate reagent

200 mg of dried leaves are hydrolysed with $H_2SO_4$ at 50□C for 30 minutes. One ml of the extract is taken and to this one ml of copper reagent A and copper reagent B are added in the ratio of 25 : 1 ml. Then it is boiled for 30 minutes and one ml of arsenomolybdate reagent is added. Deep blue colour appears slowly. The volume is made upto 25 ml and the absorbance is read at 500 nm against blank. The absorbance values are calibered from the standard graph prepare by glucose.

## 22. NON-REDUCING SUGAR

It is estimated by using the method given by Malhotra and Sarkar (1979). For this, 100 mg of dried sample is boiled in 5 ml of 1N $H_2SO_4$ is added and boil for 5 minutes in a water bath. Then it is neutralised with 1N NaOH. 1 ml of extract is taken and the remaining procedure followed is same as that for reducing sugars.

## 23. TOTAL AND FREE FATTY ACID

***Solutions***

(1) 2 : 1 chloroform : Methanol (v/v)

(2) 7% Salt solution : 7g NaCl + 100ml dd water

(3) 14% Boron Trifluoride in methanol

(4) Haxene

(5) 70 : 30 : 2 Diethyl ether : Petrolium ether : Acetic acid

(6) Chloroform

(7) Heptadecanoic acid : 0.27 g Heptadecanoic Acid (1 μmol/10 ml) 2:1 Chloroform : Methanol

***Procedure*** **:** If whole tissue is to be used homogenize it a ratio of 10:1 2:1 Chloroform : Methanol. If free fatty acids are to be measured add 10 μl of Heptadecanoic acid solution. Spin out debris at 3000 g for 10 minutes. Remove supernatant.

If using organelle or microsome pellets, grind the pellet in about 3 ml of 7% NaCl solution. Transfer to corex tube and add 5 ml 2 : 1 chloroform:methanol. If free fatty acids are to be measured add 100 μl of heptadecanoic acid solution.

Centrifuge at 10,000 g for 10 minutes and collect the lower (organic) supernatant.

***Total Fatty Acid Analysis*****:** Transfer the organic supernatant to a screwtop vial and dry completely under nitrogen.

Add 1 ml of $BF_3$ solution, blanket under nitrogen, vortex and incubate for 90 minutes at 90□C in a dry block heater.

Add 0.5 ml dd water and 1 ml hexane and vortex.

Remove hexane fraction (upper) and place in a 2 ml GC vial

*Be careful not to get water in this fraction because it will distroy the GC column.*

Dry under nitrogen to about 1/3 volume, blanket under nitrogen and analyze by GLC on a DB17 Megabore column isothermal at 160□C.

***Free Fatty Acid Analysis*** **:** Transfer the organic fraction to a test tube and dry completely under nitrogen.

Resuspend the lipids in 5 ml chloroform.

Push the extract through a Sep-Pak Silica column prewetted with 5 ml chloroform.

Ellute the column with 70 : 30 : 2 diethyl ether : Petrolium ether : Acetic acid and pool with chloroform wash.

Transfer the organic supernatant to a screw-top vial and dry completely under nitrogen.

Add 1 ml of $BF_3$ solution, blanket under nitrogen, vortex and incubate for 5 minutes at 90°C in a dry block heater.

Add 0.5 ml dd water and 1 ml hexane and vortex.

Remove hexane fraction (upper) and place in a 2 ml GC vial

*Be careful not to get water in this fraction because it will distroy the GC column.*

Dry under nitrogen to about 1/3 volume, blanket under nitrogen and analyze by GLC on a DB17 Megabore column isothermal at 160°C.

## 24. ASCORBIC ACID

*The ascorbic acid in the plants is estimated following Schafferts and Kingsley (1955) method.*

### *Reagents*

1. Trichloroacetic acid (TCA) 4% (w/v)
2. 2, 4-Dinitrophenyl hydrazene (2, 4-DNPH) – The reagent is prepared by dissolving 2 g 2, 4-DNPH in 100 ml of 9N $H_2SO_4$. This mixture is kept for overnight and filter through whatmann No. 42.
3. 85% $H_2SO_4$
4. Thiourea – The reagent is prepared by dissolving 10 g thiourea in 50 ml absolute alcohol. This solution is diluted to 100 ml with distilled water.
5. Norite

***Procedure*** **:** 500 mg of fresh leaf tissues are extracted in 24.5 ml 4% aqueous trichloroacetic acid and the homogenate is centrifused at 5000 rpm for 10 minutes. 10 ml of supernatant is taken and add 1/4 tea spoon of norite, shake for one minute and filter. 4 ml of the filtrate is reacted with 1 drop of thiourea, 1 ml of 2,4 DNPH, mix well the reaction solution. The tubes are incubated at 80°C for 10 minutes in a waterbath and later transfer to an icebath. The reaction mixture is extracted with 5 ml of 85% $H_2SO_4$ mix well and being to room temperature. The absorbance of the coloured extracts is measured at 515 nm against blank. The amounts of ascorbic acid is measured by using a standard curve.

## 25. α -TOCOPHEROL

*The α-Tocopherol in the plants is estimated following Dadlani and Agrawal.*

### *Reagents*

(1) Chloroform : Methanol (2 : 1 v/v)

(2) Ethanol

(3) α α -dipyridyl (0.5 % solution of Ethanol)

(4) Ferric chloride (0.2 % solution of Ethanol)

***Procedure :*** *(A) Extraction of Lipid* : 500 mg fresh leaf tissue is extracted in 10 ml 2:1 Chloroform : Methanol solution and homogenate is centrifuged at 10,000 rpm for 10 minutes. Two layers are seen, lower layer is carefully collect. This is the lipid extract.

*(B) Estimation of α-Tocopherol* : For drying the extract on medium, it should be carefully evaporated. The best method to achieve, this is to keep the sample in warm water (around 50□C) and blow a slow stream of nitrogen on the surface.

OR

Extract is dried under liquid nitrogen.

Now, add 1.6 ml ethanol, 0.2 ml α,α-dipyridyl and 0.2 ml of feric chloride into the dried extract. Mix well and keep in dark for 5 minutes, and intense red colour develops in reaction mixture. This reaction mixture is extracted with 4 ml of distilled water. After mixing the solution, two layers are separated out and red colour goes to aqueous layer. The absorbance of the coloured extract is measured at 520 nm against blank.

The amount of α-Tocopherol in the sample is determined using a standard curve generated using α-Tocopherol.

## 26. NUCLEIC ACID : RNA, DNA

*(1.1) Estimation of RNA* : RNA estimation is done following the Orcinol reaction (Schneider, 1945). 1 ml aliquot of extract is

taken and placed in a test tube, to which is added 1.5 ml of distilled water, 2.5 ml of concentrated HCl (Containing 0.02% $FeCl_3 . 6H_2O$) and 0.15 ml of Orcinol solution (10% in ethanol).

The reaction mixture is placed for 30 minutes in boiling waterbath to develop colour. The mixture is diluted to 7.5 ml or more with distilled water and cooled. Optical density of the solution is read at 660 nm in UV Spectrophotometer against a blank carried through the whole process.

*(1.2) Reference solution of RNA* : The reference solution is prepared by hydrolyzing commercial yeast RNA (Can be obtained from SISCO, Ranbaxy, Glaxo, etc.) in 0.3N NaOH at 40 mg/100 ml concentration.

The standard curve is prepared for the quantitative calculation of RNA in samples.

*(2.1) Estimation of DNA* : Dische's diphenylamine reaction (1955) is modified by Burton (1956) is followed for estimation of DNA. Instead of a general concentration of 1.5 g diphenylamine (DNA) per 100 ml of acetic acid, 3g of DNA is dissolved in 100 ml of glacial acetic acid as to obtain greater sensitivity while using this solution, 0.10 ml of aqeuos acetaldehyde (16 mg/ml) is added for each 20 ml of the reagent.

To the 1 ml aliquot of extract, 0.5 ml distilled water is added. 3 ml of DPA reagent is added to above solution and this mixture is incubated for 18 hours at 25 to 30□C (Kupita et al., 1961). The optical densities of the blue colour, formed on overnight incubation, are determined at 600 nm and 540 nm against blank, and the difference between the values at two optical densities are used to compute the DNA concentrations. A standard solution is with each test.

*(2.2) Reference Solution of DNA* : Sodium salt of Calf thymus DNA is used for preparing DNA standard solution. A stock solution is prepared by dissolving DNA at 0.4 μg/ml in 5 mM NaOH. From this, working standards are prepared by mixing measured volume of the stock standard with an equal volume of N-PCA and heating at 70□C for 15 minutes. Both standards are stored in the refrigerator. According to Burton (1956) no differentiation of the stock standard occurs even after six months of storage. The working standard is also used in preparing standard curve. With the help of this standard curve the quantitative values of DNA in sample are calculated and multiplied by a factor of 2 or one molecule of DNA is equal to two molecules of deoxypentose sugar.

## 27. NITRATE REDUCTASE ACTIVITY

*Nitrate Reductase Activity is measured following the method of Srivastava and Mathur (1980).*

### *Preparation of reagents*

1. 0.02 % NED N-(1-naphtyl ethylene diamine dihydro chloride) prepare in distilled water (w/v)
2. 1 % sulphanilamide (w/v) prepare in 1 N HCl
3. 0.1 % $KNO_3$ solution prepare in distilled water
4. 5 % n-propanol
5. Phosphate buffer (pH 7.4) – It is prepared by mixing 19 ml of 0.2 M $KH_2PO_4$, the mixture dilutes to 200 ml.
6. Assay Mixture :

The assay mixture is prepared as —

| | |
|---|---|
| Phosphate buffer | – 4.0 ml |
| 0.1 M $KNO_3$ | – 0.5 ml |
| 5 % n-Propanol | – 0.5 ml |
| Total volume | – 5.0 ml |

***Procedure*** : 200 mg of leaves of each sample are sliced into 2 to 3 mm fragments and placed in a light proof serum vial, in which 5 ml of assay mixture is already placed. The vials are incubated in dark at 30□C for 30 minutes.

Nitrate formed is estimated by adding 1 ml of 1 % sulphanilamide and 1 ml of 0.02 % NED to 1 ml of assay mixture after incubation. The absorbance of the pink colour developed, is measured at 540 nm in spectrophotometer. The enzyme activity is estimated in terms of nitrite (μm $NO_2$) produced per gram fresh weight of leaf tissue per hour.

$$\text{Enzyme activity} = \mu m\ NO_2\ g^{-1}\ FW\ hr^{-1}$$

The concentration of nitrate formed is calculated by calibration curve prepare by using sodium nitrate solution as a standard.

## 28. ASSAY OF NITROGENASE IN ROOT NODULES OF LEGUMINOUS PLANTS

The Nitrogenase (EC 1.19.2.1) activity in root nodules can be determined following the method of Srivastava et al. (1980). For this, root nodules are detached from the roots and 100 mg nodules are placed in a light proof 15 ml capacity serum-vial in which 0.75 ml of 100 mM Tris-HCl buffer pH 8.0 and 0.05 ml Toluene are added. The vial is then sealed with a vaccine stopper by using rubber solution. The stopper contained an abraded glass rod (2 cm. long x 4 mm diameter) freely suspended in its middle and the tip is acidified by dipping it into 1 N $H_2SO_4$. The excess acid is removed with a filter paper. The vial is evacuated and then freeze in a deep freeze (0□C) for 3 h. After thawing the content at room temperature, 0.75 ml assay mixture, which contains in a final concentration of 13.3 mM $MgCl_2$, 3.3 mM ATP and 20 mM sodium dithionite are injected. The reaction is initiated immediately by injecting 8 ml of nitrogen gas in the vial and incubation is done for 10 minutes at 30□C. The reaction is stopped by injecting 1 ml of saturated $K_2CO_3$ solution. The vials are shaken (60 strokes per minute) in an automatic shaker for 8 h for completing release of ammonia. The ammonia trappes at the tip of glass rod in the form of ammonium sulphate is determined colorimetrically with Nessler's reagent at 440 nm using spectrophotometer. The value of ammonia is calculated with standard curve drawing from ammonium sulphate.

## 29. PEROXIDASE ACTIVITY

*(A) It can be measured following Shanon et al., (1966)*

***Reagents***

1. Acetate buffer pH 5.5 (0.1M)
2. Orthodianiside 0.5 % in ethyl alcohol
3. 3% $H_2O_2$
4. 5 N $H_2SO_4$
5. 0.9% KCl

50-100 mg of fresh leaf samples are extracted for enzyme in 5 ml 0.9% KCl. Then it is centrifused at 15,000 g for 15 minutes. All the estimations are carried out in ice baths. To 50-100 μl of enzyme extract 2.5 ml of acetate buffer, 0.1 ml diansidine, 0.1 ml $H_2O_2$ are added and leave for 30 minutes. The reaction is then terminated with 0.5 ml $H_2SO_4$ (5N). The absorbance is read at 400 nm against blank. One enzyme unit is defined as change in 1 OD per minute under the assay conditions.

*(B) : It can also determine by using another method given by Maehley (1954) and George (1953).*

***Procedure*** : 200 mg of the material is taken and crush with sterilized sand in a pre-chilled pestle and mortar with 5 ml of chilled acetone in cold room at 0□C and centrifuge at 5000 g for 20 minutes. The supernatant is discarded and the pellet is redissolved in 5 ml of chilled Na-phosphate buffer (0.02 M, pH 6.4) keep for 15 minutes with occasional shaking and again centrifuge for 20 minutes. The supernatant is used for the extraction of wall bound enzyme.

## Determination of cytoplasmic peroxidase activity

It is measured by the method of Maehley (1954) and George (1953), using 20 mM guaiacol as an electron donor in presence of substrate $H_2O_2$. The reaction mixture is prepared as under—

| | |
|---|---|
| 1ml | – enzyme extract |
| 3 ml | – sodium phosphate buffer (0.02 M, pH 6.4) |
| 1 ml | – Guaiacol (20 mM) |
| 0.03 ml | – $H_2O_2$ (1 % v/v) |

The changes in absorbance is recorded at 470 nm. The activity is expressed as $\Delta A$ 470 nm $min^{-1}$ $mg^{-1}$ protein.

***Extraction of wall bound peroxidase*** : The pellet is washed four times with sodium phosphate buffer or till the washing is free of cytoplasmic peroxidase. Then this pellet is resuspended in 5 ml of 1 M NaCl solution (Ridge and Osborne, 1970) for 1 hour and centrifuge for 15 minutes at 5000 rpm. The supernatant serves as a source of wall bound peroxidase.

***Determination of wall bound peroxidase activity*** : The activity is estimated as described in cytoplasmic peroxidase following the method of Maechley (1954) and George (1953). The activity is expressed as $\Delta A$ 470 nm $min^{-1}$ $mg^{-1}$ protein.

## 30. SUPEROXIDE DISMUTASE ACTIVITY

*It can be measured by using the method given by Keith et al., 1983.*

***Reagents required***

1. 50 mM phophate buffer pH 7.0 containing 1% polyvinyl pyrrolidine.
2. 0.2 M potassium phosphate buffer pH 8.6 containing 10 EDTA and 2 x 10 M cytochrome C (substrate).
3. Alkaline dimethylsulfoide (DMSO containing 5 mM NaOH).

***Method*** **:** 200 mg of the leaf tissue is extracted in 50 mM phosphate buffer (3 ml) and is centrifuged at 30,000 g for 30 minutes. To the 0.2 ml of the extract one ml of the substrate is added and keep for 30 minutes. Then the reaction with 0.5 ml alkaline DMSO. The blank is terminated with DMSO containing no NaOH. The absorbance of the reduced cytochrome is measured at 550 nm against samples prepare under similar conditions except that DMSO do not contain NaOH. A unit of enzymatic activity is the amount of enzyme which causes 50% inhibition of alkaline DMSO mediated cytochrome C reduction.

## 31. SULPHATE CONTENT

0.2 g dried ground leaf sample is taken in a 50 ml Kjeldahl flask and is digested with a mixture of 0.1 g selenium oxide, 10 ml of conc. $HNO_3$ and one ml of conc. HCl. To the residue 0.2 ml of dilute HCl is added and again evaporated to dryness. One ml conc. HCl and 25 ml distilled water is added and then boil and filter in 100 ml volumetric flask. The filtrate is made to 50 ml by 5 ml 3 % glycerol and distilled water.

10 ml of aliquot is taken in a test tube and 0.5 ml of 2% Barium chloride solution is added. Sulphate is precipitated as Barium sulphate and the absorbance is read at 420 nm against blank. The actual sulphate content is calibrated from the standard curve prepare by potassium sulphate solution and is expressed as mg/g dry wt.

## 32. TOTAL NITROGEN

This is estimated by using the method given by Mc Donald (1978). 10 g of oven dried plants material is digested in corning test tube with 12 drops of con. $H_2SO_4$ and a pinch of digestion mixture, which consists of following chemicals

| | |
|---|---|
| Potassium sulphate | 2.0 g |
| Sodium thio sulphate | 1.5 g |
| Salicylic acid | 1.5 g |
| Selinium | 1.0 g |
| Total | 5.0 g |

After cooling the digestion mixture add a few drops of hydrogen peroxide and heat for some time. Finally the colourless digested fluid is made up by distilled water to 10 ml volume. Then to this 0.2 ml of sodium silicate (10%) and sodiumhydroxide (10%) are added and shake thoroughly. The final volume is made to 10 ml with distilled water. The 2 ml of diluted (1:2) Nessler's reagent is mixed and shake. Then after 10 minutes the absorbance of yellow is read at 425 nm. The amount of nitrogen is calculated with the help of standard curve prepare from ammonium sulphate. The results are expressed as percentage.

## 33. SULPHURDIOXIDE GENERATION

The gaseous pollutant, sulphurdioxide is generated by bubbling air through 1% aqueous sodium metabisulphite solution at a fixed rate. The air flow rate is first standardized in a 1 x 1 x 1 m polythene open top chamber. The generated gas is then measured at one hour interval with Toxic gas monitor. The avearage value obtained is considered as the concentration of the gas to which the plants are exposed. The amount of the gas generated is dependent on the air flow rate.

## 34. TDF (Total Dust Fall)

The total dust fall is estimated with the help of uniform size jars. Five jars are placed at each site at an height of 8 meters from ground surface, open to sky for two months. They are then brought to the laboratory and settled dust is weighed and express as total dust fall g/m$^2$/day.

## 35. LEAF WASH pH AND LEAF EXTRACT pH

Leaves are washed in 50 ml of distilled water twice and the pH of the water is determined by Digital pH meter directly. For estimating the Leaf extract pH, homogenise 5g fresh leaves in a mortar with 25 ml of double distilled water and measure the pH of homogenite directly by Digital pH meter.

## 36. TGM STUDY (Toxic Gas Monitor)

***Principle*** : The Toxic Gas Monitor measures the concentration of a particular gas in the air by monitoring the amount of colour change produced in the analyzers reagent systems. Air to be analysed is continuously drawn into the instrument via an internal vaccum pump and is made to pass through proper absorbing reagent via series of tubings and the colorimetric estimation of the gas is displayed on the digital read out.

***Procedure*** : *(i) Sulphurdioxide* : Suphur dioxide estimation in the embient air is done by continuously absorbing the air at an fixed rate. The air sucked by the pump is then reacted with pararosaniline and formaldehyde reagent to form intensly coloured pararosaniline methyl sulphonic acid, whose intensity is measured at 550 nm. The intensity of the colour produced is proportional to the concentration of $SO_2$ absorbed.

*(ii) Nitrogendioxide* : The air is absorbed into the azodye forming reagent with chromium trioxide to convert NO present to $NO_2$. The intensity of the azodye formed is measured at 550 nm.

The digital read out for both the gaseous pollutants are calibrated from the standard graph supplied with the monitor.

## 37. HVS STUDY (High Volume Sampler)

***Principle*** : Air at a predetermined rate is sucked in by a cyclone (motor) through a filter and there by making the suspended particulates in ambient air adsorb to the filter. With initial and final weight of the filter paper the amount of suspended particulate is estimated in the total volume of the air sucked in.

***Procedure*** : Air at 0.5 LPM rate is sucked in by the motor through a 250 x 200 mm, 0.3 micron glass fibre filters. The initial air flow rate and weight of the filter paper is noted. The sampler is run for 8 hours at each site. Final air flow rate and weight of filter paper is noted. Mass concentration of SPM is evaluated by the formula

$$\text{SPM} = \frac{[\text{Wf} - \text{Wi}]}{\text{V}} \times 10$$

Where, Wi and Wf are the initial and final weight of the filter paper (g) and V is the air volume in cubic meters.

## 38. CPS STUDY (Cascade Particle Separator)

***Principle*** : The principle remains same as that of the high volume sampler, only the different sized suspended particulates flow from ring nozzle of stage and are separated by crushing and attacking on the surface by impactor method.

***Procedure*** : Filter papers with differential retention capacity are placed separately on different ring nozzle of stage. Air is sucked in at the flow rate of LPM through these filter papers. The sampling is done for 8-24 hours at various sites. The various particle fractions that can be measured are, <0.7, 0.7-1.6, 1.6-5.4, 5.4-10.9 and >10.9 μ.

## 39. AAS STUDY (Atomic Absorption Spectrophotometer)

***Principle*** : Light at the resonance wavelength of initial intensity $I_0$ is foccused on the flame cell containing ground state atoms. The initial light intensity is decreased by an amount determined by the atom concentration in the flame cell. The light is then directed on to the detector where the reduced intensity "I" is measured. The amount of light absorbed is determined by comparing I to $I_0$.

Hollow cathode lamps are used as light source as they are designed to emit the atomic spectrum of a particular element, and the specific lamps are selected for use depending on the element to be determined.

The specific limitations and conditions for each element are as follows :

| Element | Wavelength (nm) | Slit width (nm) | Flame Gases | Sensitivity check @ |
|---|---|---|---|---|
| Cadmium | 228.8 | 0.7 | A–Ac | 1.5 |
| Chromium | 357.9 | 0.7 | A–Ac | 4.0 |
| Cobalt | 240.7 | 0.2 | A–Ac | 7.0 |
| Copper | 324.8 | 0.7 | A–Ac | 4.0 |
| Iron | 248.3 | 0.2 | A–Ac | 5.0 |
| Lead | 283.3 | 0.7 | A–Ac | 20.0 |
| Manganese | 279.5 | 0.2 | A–Ac | 2.5 |
| Nickel | 232.0 | 0.2 | A–Ac | 7.0 |

A – Ac = Air – Acetylene gas; @ = Metal concentration (mg/l) in aqueous solution which will give a reading of approximately 0.2 absorbance units.

## 40. METHODOLOGY FOR TISSUE CULTURE

### "*IN VITRO CULTURE*" TECHNOLOGY

*In vitro culture,* in which isolated organs, tissues, cells and even protoplasts (cells without wall) of plants are grown in nutrient medium in laboratory under aseptic conditions, is one of the growth areas of biotechnology.

**Types of *In vitro Culture* of Plants**

*(a) Culture of (almost) whole plants*

*(i) Seed Culture* : A seed may be sown in vitro from which a seedling, and finally, a plant developed, e.g. Orchids.

*(ii) Embryo Culture* : An excised embryo is grown.

*(b) Organ Culture*

An isolated organ is grown. Different types can be distinguished, e.g. anther culture, ovary culture, nucellus culture.

*(c) Tissue Culture*

A differential tissue is isolated and grown.

*(d) Single cell Culture*

The growing of individual cells which have been isolated from a tissue, callus or suspension culture.

*(e) Protoplast Culture*

The culture of protoplasts (cells without outer wall).

Apart from the actual subject matter of "in vitro culture" technology of (almost) whole plant, organ, tissue, cell and protoplast culture. There are three very important aspects which deserve mention first. These aspects are *nutrient media* maintenance of *aseptic condotions* and *aeration*.

### NUTRIENT MEDIA

Nutritional requirements for optimal growth of plant organ, tissue, cell and protoplast *in vitro* generally vary from species to species. Even tissues from different parts of a plant may have different requirements for satisfactory growth. It is, therefore, no single medium as such can be suggested as being entirely satisfactory for all types of *in vitro* cultures. When starting with a new system, it looks essential to work out a medium that would fulfil the specific requirements of that system. This is why many recipes of nutrient media have been developed (Table 1).

Some of the earliest *in vitro* culture media, e.g. root culture medium of White (1943) and callus culture medium of Gautheret (1939) were developed from nutrient solutions earlier used for whole plant cultuere. White evolved the medium from Uspenski and Uspenskaia's medium (1925) for algae, and Gautheret's medium is based on Knop's (1865) salt soluton. All subsequent media for mutations are based on White's and Gautheret's media.

In order to formulate a suitable medium for a new system, it would be better to proceed with well-known basal medium such as MS (Murashige-Skoog) or $B_5$ (Gamborg et al.). By making minor qualitative and quantitative changes, through a series of experiments, a new medium may be evolved to accommodate specific requirements of the plant materials in question.

## Preparation of Media

*1. Stock - A :* In a conical flask, 200 ml of distilled water is taken. All the macronutrients are weighed and are added serially till it completely dissolves. Similarly, in another conical flask containing 200 ml distilled water, all the micronutrients are weighed and are dissolved serially. Finally make up volume to 500 ml.

*2. Stock - B :* Similarly, all organic compound (vitamins, auxins, cytokinins) are added as desired. About 3% sucrose is added. Finally the volume is made upto 500 ml.

3. Add both the stock (i.e. Stock A & B) and a clear and transparent solution is obtained. pH is adjusted using NaOH, HCl or KOH.

4. Finally agar is added of about 0.8% and is heated until the clearity of the solution is obtained.

5. The prepared culture medium is poured in a culture tubes.

6. Medium is sterilized in a validated autoclave at 1 kg/cm (15 psi) 121□C, for the time period according to the volume.

Table 1. Media Amount (in mg/l)

| Component | White (1943) | Heller (1965) | Murashige & Skoog (1962) | Gamberg *et al.* (1969) | Nagata & Takebe (1971) | Nitsch (1969) |
|---|---|---|---|---|---|---|
| **(A) Macronutrient** | | | | | | |
| KCl | 65 | 750 | – | – | – | – |
| $(NH_4)_2SO_4$ | – | – | – | 134 | – | – |
| $NH_4NO_3$ | – | – | 1650 | – | 825 | 720 |
| $NaNO_3$ | – | 680 | – | – | – | – |
| $KNO_3$ | 80 | – | 1900 | 2500 | 950 | 950 |
| $Ca(NO_3)_2$ | 300 | – | – | – | – | – |
| $CaCl_2.2H_2O$ | – | 75 | 440 | 150 | 220 | – |
| $MgSO_4.7H_2O$ | 720 | 250 | 370 | 250 | 1233 | 185 |
| $Na_2SO_4.10H_2O$ | 200 | – | – | – | – | – |
| $KH_2PO_4$ | – | 125 | 170 | – | 680 | 68 |
| $NaH_2PO_4.7H_2O$ | 16.5 | 150 | – | 150 | 1233 | – |
| $CaCl_2$ | – | – | – | – | – | 166 |
| **(B) Iron** | | | | | | |
| $Na_2EDTA$ | – | – | 37.3 | 37.3 | 37.3 | 37.3 |
| $FeSO_4.7H_2O$ | – | – | 27.8 | 27.8 | 27.3 | 27.8 |
| $FeCl_3.6H_2O$ | – | 1.0 | – | – | – | – |
| $Fe_2(SO_4)_3$ | 2.5 | – | – | – | – | – |

| Component | White (1943) | Heller (1965) | Murashige & Skoog (1962) | Gamberg *et al.* (1969) | Nagata & Takebe (1971) | Nitsch (1969) |
|---|---|---|---|---|---|---|
| **(C) Micronutrients** | | | | | | |
| $MnSO_4.4H_2O$ | 7 | 0.01 | 22.3 | | 22.3 | 2.5 |
| $MnSO_4.H_2O$ | | | | 10 | | |
| $ZnSO_4.7H_2O$ | 3 | 1 | 8.6 | 2 | 8.6 | 10 |
| $H_3BO_3$ | 1.5 | 1 | 6.2 | 3 | 6.2 | 10 |
| KI | 0.75 | 0.01 | 0.83 | 0.75 | 0.83 | – |
| $Na_2MoO_4.2H_2O$ | – | – | 0.25 | 0.25 | 0.25 | 0.25 |
| $CuSO_4.5H_2O$ | – | 0.03 | 0.025 | 0.025 | 0.025 | 0.025 |
| $CoCl_2.6H_2O$ | – | – | 0.025 | 0.025 | – | – |
| $CoSO_4.7H_2O$ | – | – | – | – | 0.03 | – |
| $NiCl_2.6H_2O$ | – | 0.03 | – | – | – | – |
| $AlCl_3$ | – | 0.03 | – | – | – | – |
| **(D) Organic Compound** | | | | | | |
| Biotin | | | | | | 0.05 |
| Meso-inositol | 100 | 100 | 100 | 100 | 100 | 100 |
| Nicotinic acid | 0.5 | – | 0.5 | 1.0 | – | 5 |
| Pyridonine-HCl | 0.5 | – | 0.5 | 1.0 | – | 0.5 |
| Thiamine-HCl | 0.1 | 1.0 | 1.0 | 10.0 | 1.0 | 0.5 |
| Glycine | 30 | – | 2.0 | – | – | 2.0 |
| Ca-D-pantothenate | 1.0 | – | – | – | – | – |

| Component | White (1943) | Heller (1965) | Murashige & Skoog (1962) | Gamberg *et al.* (1969) | Nagata & Takebe (1971) | Nitsch (1969) |
|---|---|---|---|---|---|---|
| Folic acid | – | – | – | – | – | 0.5 |
| **(E) Carbohydrate** | | | | | | |
| Sucrose | 2% | 2% | 3% | 2% | 1% | 2% |
| Mannitol | – | – | – | – | 13%w/v | – |
| **(F) Growth Regulator** | | | | | | |
| Kinetin | – | 0.25 | 0.04-10 | 1.0 | – | – |
| 2,4-D | 6.0 | 1.0 | – | – | – | – |
| IAA | – | – | 1-30 | – | – | – |
| NAA | – | – | – | – | 3.0 | – |
| 6-Benzyl-aminopurine | – | – | – | – | 1.0 | – |
| pH | 5.5 | – | 5.7-5.8 | 5.5 | 5.8 | – |

| Volume (ml) | Minimum Sterilization time at 121□C |
|---|---|
| 20-50 | 15 min. |
| 75 | 20 min. |
| 250-500 | 25 min. |
| 1000 | 30 min. |
| 2000 | 40 min. |

7. Medium is allowed to cool and it gets solidify (if an agar medium).

## Preparation of Stock solutions of Plant Growth Regulators

1. Stock solutions of plant growth regulators are prepared by first dissolving them in a few drops of 1N NaOH or 1N HCl and then making up the volume with distilled water.
2. mg/ml stock solutions of plant growth regulators is prepared by dissolving 50 mg in 50 ml of distilled water and is stored in refrigerator.
3. For using micromole concentrations, it is convenient to make a 10 M solution of which 1 ml is added to the 1 l of medium and will give 1 μM concentration of the compound.

For e.g. to make 10 M stock of IAA (M.W. 175) 17.5 mg is weighed and is dissolved in a volumetric flask of 100 ml, in a few drops of 1N NaOH or ethanol. Finally the volume is made up with distilled water and is stored in refrigerator.

4. Among plant growth regulators, IAA is protected from light by covering the bottle with black paper as it degrades in light.
5. All stock solutions must be used within four weeks of preparation.
6. Solution should be discarded if it gets precipitated or there is any colour.

## ASEPTIC CONDITIONS

The nutrient media used in *in vitro* culture studies, specially when they contain sugar, attract variety of microorganisms and help them luxuriently growing in it. The microorganisms grow at the faster rate than the cultured tissue in medium and finally, kill the latter. It is, therefore, obligatory that a completely aseptic condition is maintained in and around the culture equipments so that the microorganisms fail to contaminate the nutrient media.

Following of the three main sources of contamination of the media, and the subsequent methods to overcome them are –

1. The microorganisms may be present in the nutrient medium right from the beginning. These microorganisms are destroyed by properly plugging and autoclaving (steam heating under pressure). The medium can be completely sterilized by maintaining it at 121□C for the time period according to volume.
2. Finally, strict precautions are taken to prevent the entry of microorganisms when the plug of a culture is removed to transfer the plant material to the nutrient medium or from one medium to the other. For this, all operations are done in completely aseptic conditions.

## SURFACE STERILIZATION

The plant part to be cultured may carry microorganisms with it. To avoid it, the plant part is surface sterilized as bacterial and fungal growth are detrimental to culture. Explants are washed in mild detergent before treatment with the disinfactory solution. they are rinsed thoroughly under running tap water for 15-30 min. and is submerged into the disinfectant solution and is agitated gently. Disinfectant solution commonly used are (a) Calcium hypochloride (9-10%) for 15-30 min. and (b) Mercury chloride (0.1-1%) for 5-10 min.

Sterilization is enhanced by placing the material in a 70% ethyl alcohol solution prior to treatment with another disinfectant solution. The use of two steps (two sources) sterilization procedure is of more beneficial. A wetting agent such as Tween-20 or 80 is used to reduce surface tension and allow better surface content. It is added along with the disinfectants under sterile conditions, the solution is decant and explants are rinsed with sterile distilled water several times. If it is conducted under vaccum, this results in the removal of air bubbles and provides a more efficient sterilization proccess.

## INNOCULATION

Air flow system of laminar is switched on for about 1 hr and UV lamp for about 15 min. before innoculation. Hand should be washed with soap and water. The surface of the laminar bench should be wiped with a swab of spirit. All the requirements such as spirit lamp, forceps/scapel in a beaker half filled with spirit, surface sterilized material, sterile petridishes, sterile distilled water and sterile medium, all are arranged on a bench according to the convenience in working. The forceps and scalpel blades are flamed and are kept for cooling on a cooling rack or sterile petridish. Plant material is then cut into pieces of desired size in a sterile petridish. The tube or flask is opened near the flame and mouth of the tube is heated for sometime so that viral or bacterial contamination can be avoided. The cut explant are innoculated on the medium with the help of forcep and is pressed lightly and plug is replaced. All the steps are carried out quickly to avoid the contamination.

### AERATION

While *in vitro* culturing is going on the cultured plant part should be given proper aeration. If the plant part is grown on the surface of a semi-solid medium it acquires sufficient aeration without a special device. However, if liquid medium is used the plant part, since it gets submerged, requires some special device for proper aeration. One device is to use a "filter paper bridge" whose two legs remain dipping in the medium, and the horizontal part carrying the plant part is raised above the level of the medium. Agetation of the medium by passing through it filter sterilized air also serves the purpose. More commonly, however, aeration is provided by shaking the flask or tubes on an automatic shakes. Besides supplying 0 to the plant part, shaking of the medium also brings about dissociation of the tissue into small clumps and even single cells. The importance of single cells in the cloning and higher plant genetics has been greatly appreciated.

## 41. MEASUREMENT OF METAL TOLERANCE AND PHYTOTOXICITY

Every organisms, (Plants or Animals), regardless of whether it lives in a metal-enriched environment or not, has certain ability to cope with non-essential metals or excessively available essential metals, although, there are limits in the occurrence of metals tolerance in a lot of higher plants. Several workers have referred to this ability as "tolerance" or "resistance". It is to restrict these terms to cases of a heritable increase in the ability to cope with excessive metal, usually naturally or artificially selected under the pressure of a toxic level of metal exposure. Such heritable tolerances occur in a limited number of plant species and have been demonstrated for Zn, Pb, Cu, As, Cd and other metals. Evolution of tolerant population may occur as rapidly as within 5 to 10 years, or more slowly dependent on the selection pressure prevailing.

Tolerance to metals can be demonstrated at different levels of integration. Basically metal tolerance is a constitutive property, present in every living cell, tissue and organ of the plant and animals. It is maintained in cell tissue cultures prepared from tolerant plants. Therefore, tolerance can be tested using different tissue, organs or whole organisms in case of plant. Although, it can not be excluded that process taking place at integration levels higher than the cell may substantially contribute to the tolerance of the whole plant.

*There are four methods for the measurement of tolerance and phyto-toxicity in plant systems :*

1. Comparative plasmology
2. Rooting technique
3. Whole plant testing
4. Chlorophyll synthesis

**Method- I :** *Comparative plasmology*

First time Repp (1983) elaborated a test for plasma membrane integrity after exposure to toxic metals. Sections of leaf or stem tissue, between 2 to 4 cell layer thick, are incubated are series of metal concentrations, either nutrient addition or a nutrient solution. After 5 to 6 hours (in some cases 24 hours) the cells ability to plasmolyse or deplasmolyse is tested using 1 M glucose.

OR

After keeping the test materials in metal solution nearly 5 to 6 hours solute leakage will be measured by any sensitive conductimeter. The solute leakage will expressed as the difference between the initial and final conductivity of the solution.

*The advantage of this technique is that it allows testing of established plants in field and in the laboratory without cloning them.*

**Method - II :** *Rooting Technique*

Roots are more directly confronted with toxic metals in the environment than shoots, except in the case of aerial metal deposition. Therefore, root growth usually responds more rapidly to metal exposure than shoot growth. This is probably the reason that root growth is the most widely used parameter in metal tolerance tests. A commonly used measure of tolerance is the tolerance index.

$$\text{Tolerance Index} = \frac{\text{Root growth in the metal solution}}{\text{Root growth in control solution}}$$

The use of tolerance index certainly allows a perception of the coarse patterns of interspecific and intraspecific variation in metal tolerance. The resolution of this index, however is limited and depends on the metal concentration chosen for testing, as well as the range of variation in tolerance present among the plants to be tested.

Moreover, low metal concentrations may stimulate root growth, specially in case of tolerant plant.

**Method III :** *Whole plant testing*

Several workers have used the relative growth rate of the whole plant as an effect parameter in metal tolerance tests. In view of the fact that tolerant and non-tolerant plants differ may differ in the rate of translocation of metals in the roots. It is conceivable that they may also differ in the degree of shoot growth inhibition, relative to the degree of root growth inhibition, if they exposed to toxic metal concentration. If so, then whole plant testing may be expected to differ in many attributes from those obtained with a rooting test.

**Method IV :** *Chlorophyll synthesis*

This is one of the best method for the measurement of the toxicity and tolerance of plant in relation to heavy metals. Photosynthesis is an important process of plant in which light energy is converted into chemical energy. The chlorophylls have long been recognized as the primary light acceptors in plants and they are invariably present in every organism which carries out photosynthesis with absorption of carbondioxide and evolution of molecular oxygen. Heavy metals inhibit the photosynthesis in various ways. Stunting and chlorosis are commonly observed reactions of metal toxicity. One of the commonly reported symptoms of Cadmium toxicity is the chlorosis of leaves. It has been proved that any concentrations of toxic metals showed phytotoxicity to the plants.

# PART – II
# METHODOLOGY FOR WATER ANALYSIS

## PART – II

# METHODOLOGY FOR WATER ANALYSIS

## 1. TEMPERATURE

**(i) *General*** : Temperature plays an important role in biological activities, fish life, limnological studies etc. Temperature governs solubility of oxygen, carbon dioxide, bicarbonate carbonate equilibrium.

**(ii)*Apparatus*** : A thermometer having a quick response and having atleast divisions for 0.5°C.

**(iii) *Procedure*** :

(a) *Air temperature* (ambient temperature) Hold a dry thermometer at least one meter above ground, keeping it away from any heat source or direct solar radiation and read.

(b) *Water Temperature* : Thermometer is dipped directly into water body and a constant reading is noted. Temperature can be measured in the sampling bottle immediately after taking the sample. Ensure that container attains the temperature of the water sample.

**(iv)*Expressions of results*** : Temperature may be recorded to the nearest of 0.5°C.

**(v) *Precautions*** : While taking temperature of water sample ensure that the mercury bulb of the thermometer is dipped in the water. Before making any temperature measurement, wipe the thermometer with tissue paper or filter paper. Whenever the embient air and water temperature both are to be measured, it is suggested that the earlier should be followed by the later.

## 2. pH VALUE

**(i) *General* :** pH is the negative logarithm of hydrogen ion concentration (activity of hydrogen ion) in moles per litre at a given temperature. The pH of natural water is controlled by the carbon dioxide/bicarbonate equilibria and usually ranges from 1.0 to 9.0. The majority of water are slightly basic due to the presence of bicarbonates and carbonates. pH affects the chemical and biological properties of liquids and hence is an important estimation; e.g. for controlling corrosion, and water and waste treatment processes.

Electrometric and colorimetric methods are available for the measurement of pH.

### Electrometric method

**(i) *Principle*** : The pH is determined by measurement of the electromotive force of a cell comprising an indicator electrode, an electrode responsive to hydrogen ions (glass electrode) immersed in the test solution, and a reference electrode (usually colomelelectrode). Contact between the test solution and the reference electrode is achieved by means of a liquid junction, in the reference electrode. A difference of 1 pH unit produces a potential charge of 58.16 mV at 25□C. The electromotive force is measured with a pH to read directly as pH value.

*Interferences :* The glass electrode is free from interferences such as colour, turbidity, collodial matter, oxidant reductants. In case, salinity is high a sodium ion error at pH 10 and above has to be compensated. Temperature affects the potential at electrodes and the ionization in the sample. The first effect can be overcome by a temperature compensation adjustment provided on the most of the instruments. The second effect is inherent inthe sample and is taken in to consideration by recording both temperature and pH of each sample.

**(ii) *Apparatus* :**

(a) *pH Meter* – Mains or battery operated model (for field measurement) with temperature compensation adjustment and reading to an accuracy of 0.1 pH units.

(b) *Glass electrode* – This must be compatiable with the pH meter used and should cover the entire pH range with minimum sodium ion error.

(c) *Reference electrode* – Calomel electrode is widely used. Less concentrated solutions of potassium chloride (e.g. 3.5 M KCl or 350 g/l) are more satisfactory than the saturated solution usually used which very often leads in clogging of the electrode.

**(iii) *Reagents*** : Standard pH buffer solutions : This can be prepared from buffer tablets which are commercially available. However, an analyst can prepare his own solutions by referring to standard text book.

**(iv)*Procedure* :**

(a) Rinse the electrodes with distilled water and dry by gentle wiping with a soft tissue paper.

(b) Standarize the instrument with the electrode immersed in a buffer solution with a pH close to that of the water to be tested.

(c) If manual temperature compensation is provided, set it to the temperature of the buffer.

(d) Check the electrode response occasionally by measuring a second buffer solution of different pH.

(e) Rise with distilled water and dry with tissue paper after every measurement. The difference in reading for the standard buffers should not exceed 0.1.

(f) Immerse the electrode in the sample.

(g) Set the temperature control adjustment knob to the temperature of the sample. Taking final reading after 30 seconds.

**(v) *Expressions of Results*** *:* pH value should be reported to the nearest 0.1 pH unit. The temperature of the sample at the time of the pH measurement should be indicated.

**(vi)*Precision and Accuracy*** : The precision and accuracy attainable with a pH meter depends on the type and condition of the instrument. A precision of ± 0.02 pH unit and an accuracy of ± 0.05 pH unit can be achieved. However, ± 0.1 pH is satisfactory. Hence, pH value should be reported to the nearest of 0.1 pH unit.

**(vii)*Precautions :***

(a) Before making any measurement rinse the

electrodes with distilled water and dry it with the tissue paper.

(b) Ensure adequate level of filling solution in the reference electrode.

(c) Always keep the electrode in distilled water, never allow them to dry out.

(d) Standarize the instrument in case of any doubts with the pH buffer closer to the pH of the test sample.

(e) Adjust the temperature compensation knob before reading.

(f) After immersing the electrode, wait for sometime so as to stablize the system. Take the reading as soon as it becomes constant.

(g) As for as possible it should be measured at the site or as soon as possible.

(h) No chemical preservation of sample is permissible.

(i) Universal glass electrodes ensure good results.

## 3. ELECTRICAL CONDUCTIVITY

**(i) *General*** : Conductivity is a measure of the ability of a water sample to carry an electrical current and it depends on the ionic strength of the water. This mostly depends upon the nature of the various dissolved ionized substance, their actual and relative concentrations and the temperature at which the measurement is made.

Most of the inorganic acids, bases and salts (such as hydrochloric acid, sodium carbonate and sodium chloride) are relatively good conductors. Conversely, most organic compounds, such as sucrose and benezene that do not dissociate in aqouous solution conduct a current very poorly, if at all.

The determination of electrical conductivity is a rapid and convenient means of estimating the concentration of electrolytes. In waters containing mostly mineral salts (waters used for public supply and many other ground or surface waters) this will not be very different from that of the dissolved solids. Hence, in such a samples conductivity can be used as a measure of dissolved solids.

For a particular samples one can have a factor to convert electrical conductivity to dissolved solids. This factor will very from sample to sample. Thus, in some cases measuring the conductivity and converting it into dissolved solids, can be used as a check for the quality of the effluent discharged or for monitoring the treatment plant. The factor usually varies from 0.55 to 0.9. The factor is multiplied by conductivity (μmhos/cm) to obtain dissolved solids. Some other important applications of conductivity measurements are :

(a) The purity of distilled and deionized water can be checked by conductivity determination.

(b) To check the variations in the dissolved mineral concentration of raw water or waste water samples quickly.

(c) Determination of the amount of ionic reagent required for certain precipitation and neutralization reactions, the end point in such cases is determined by a sudden change in the slope of the conductivity curve.

(d) An approximate estimation of the eliquots which may be taken for the chemical determination.

(e) Since it is a measure of dissolved solids it can be used to assess the effect of diverse ions on chemical equilibris, physiological effects on plant or animal, corrosion rates etc.

The standard unit of electrical conductivity is the Siemen per meter ($Sm^{-1}$) (mhos $m^{-1}$) in order to avoid the expression of results in small decimal fractions, a smaller unit the millisiemen per meter (ms $m^{-1}$) is generally used.

Freshly, prepared distilled water has a conductivity of 0.1 to 0.2 ms $m^{-1}$ (or 1 to 2 μmhos $cm^{-1}$) or even less, but this increases on storing due to absorption of atmospheric carbon dioxide to the extent of 0.2 to 0.4 ms $m^{-1}$ or 2 to 4 μ mhos $cm^{-1}$ in about a weeks time. The electrical conductivity of most of the fresh and finished waters is in the range of 5 to 50 ms $m^{-1}$ (or 50 to 500 mhos $cm^{-1}$). The values for highly mineralized waters go up to 100 ms $m^{-4}$ and even higher. Some industrial wastes may have conductivities even more than 1000 ms $m^{-1}$.

Electrolytic conductivity increases with temperature at a rate of approximately 2% degree C. Thus, significant errors can result from inaccurate temperature measurements at 25□C.

No chemical preservation of samples is permissible.

**(ii) *Apparatus — Conductometer***

Many instruments of different ranges are available, most of which are mains operated. Essentially they consist of following the main parts.

(a) *Conductivity cell :* This consists of a pair of rigidly mounted electrodes. Each cell has its own constant depending on its shape and the position and size of the electrodes. This constant is required to calculate the conductivity of the sample. This can be determined by using a standard solution (e.g. 0.01 N KCl use Table 1 in the conductivity water or by comparison with a cell of accurately known constant).

(b) *Instrument for measuring electrical conductivity* : Most of the instruments consist of a source of alternating current, a wheatstone bridge and a null indicator. Other instruments measure the relation of alternating current through the cell, a voltage across it and have the advantage of a linear reading of conductance. The instrument chosen should be capable of measuring conductivity with an errors not exceeding 1% of 0.1 $Sm^{-1}$ (or 1 $\mu$ mhos $cm^{-1}$) whichever is greater.

The conductivity cells should be stored in distilled water when not in use. Cells with plantinium electrodes should be repletinized whenever it gives ornatic readings.

To platinize, prepare a solution of 1 g chloropletinic acid (platinium chloride) and 12 mg lead acetate in 100 ml water. Immerce the electrodes in this solution and connect both to the negative terminal of a 1.5 V dry cell battery. Connect the positive side of the battery to a piece of platinium wire and dip the wire into the solution. The amount of current should be such that only a small quantity of gas is evolved. Continue the electrolysis until both cell electrodes are coated with platinium black. The platinizing solution may be saved for subsequest use. The electrodes should be rinsed thoroughly and when not in use keep them immersed in distilled water.

A different type of electrode with a bright surface is also available. Such cells consists of electrodes made from durable metals like stainless steel are widely used for continuous monitoring and field studies. These cells are calibrated by comparing the conductivity of the water being tested with the results obtained with a laboratory instrument.

*Thermometer* : Capable of reading to the nearest of 0.1□C and covering the range 20 to 30□C.

Table 1. Conductance of Potassium chloride solution at 25□C.

| Concentration (N) | Conductance (ms $m^{-1}$) | |
|---|---|---|
| | Equalant | Specific |
| 0.0001 | 14,943 | 14.94 |
| 0.001 | 14.695 | 147 |
| 0.01 | 14.127 | 1,413 |
| 0.1 | 12,896 | 12,980 |
| 1.0 | 11,187 | 111,900 |

***(iii) Reagents***

Conductivity water is used for the preparation of all the standard solutions and this should have a very low conductivity. This must be completely free from carbon dioxide. Reaistilled water must be boiled just before use and allowed to Cool. The conductivity of this water should be less than 0.1 ms $m^{-1}$.

Standard potassium chloride, 0.01 N : Dissolve 745.6 mg anhydrous KCl in conductivity water and make upto 1000 ml at 25□C. This is the standard reference solution having specific conductance of 1,41.3 $ms^{-1}$. It is satisfactory for most waters when using a cell with a constant between 1 and 2. For other cell constants, stronger or weaker potassium chloride solution listed in Table 1 will be needed. Store this solution in glass stoppared pyrex bottle.

***(iv) Procedure***

(a) *Determination of cell constant* : Rinse the conductivity cell with at least three portions of 0.01 N KCl solution. Adjust the temperature of a fourth portion to 25.0 ± 0.1□C. Measure the resistance of this portion and note the temperature. Compute the cell constant C from equation.

$$C = \frac{0.001413\ RKC1}{1 + 0.0200\ (t - 25)}$$

(b) *Conductivity measurements* : Rinse the cell with distilled water followed by portion of sample to be tested. Adjust the temperature of the sample to 25 ± 0.1□C. Measure the resistance of the sample and note the temperature.

(c) *Calculations* : The electrical conductivity of the sample can be obtained by multiplying the observed conductivity (in ms $m^{-1}$) by the cell content. If the instrument measure electrical resistance then divide cell constant by observed electrical resistance.

$$E = C - Em \text{ or } E = \frac{C}{RM}$$

To account for the temperature differences, divide the result by the factor [1 + 0.02000 (T – 25)]ic.

$$E = \frac{C \times Em}{1 + 0.02000\,(T - 25)} \text{ or } E = \frac{C}{Rm}\ \frac{1}{1 + 0.0200\,(T - 25)}$$

Where;

E = Electrical conductivity of sample in ms $m^{-1}$

C = Cell constant

Em = Electrical conductivity measured ms $m^{-1}$

T = Sample temperature (□C)

Rm = Electrical resistance measured (K ohm.m)

### *(v) Expression of results*

Express the result in millisiemen per meter ($msm^{-1}$). Round off the result as shown below :

| Range ms $m^{-1}$ | Record to nearest |
|---|---|
| 1.0 to 2.0 | 0.02 |
| 2.0 to 5.0 | 0.05 |
| 5.0 to 10.0 | 0.1 |
| 10.0 to 20.0 | 0.2 |
| 20.0 to 50.0 | 0.5 |
| 50 onwards | To nearest whole number. |

### *(vi) Precision and accuracy*

The precision and accuracy of measurement depends on the instrument used and accuracy with which cell constant is determined. The precision and accuracy of about 12% is expected with a satisfactory equipment and a good analyst.

*Note* : Several conductometers indicate reading as us $cm^{-1}$ or μmho $cm^{-1}$. This reading is to be divided by 10 to convert it into $msm^{-1}$. Following conversions may used :

$$10\ \mu s\ cm^{-1} = 40\ \mu\ mho\ cm^{-1} = msm^{-1}$$

***(vii) Precautions***

(a) When electrodes are not being used always keep them in water.

(b) Always record the temperature and apply temperature correction. Error in this may cause significant error in the results.

(c) The final conductivity results mainly depend on the cell constant, and hence cell constant determination will cause error in conductivity measurement also.

(d) Always use conductivity water for preparing standards.

(e) Conductivity measurements should be made on the spot or as soon as possible.

## 4. HARDNESS

Hardness in water is mainly due to calcium and magnesium ions. Though some other polyvalent ions such as aluminium, iron, manganese, strontium and zinc also contribute to hardness. In natural waters it is defined as sum of the calcium and magnesium ions expressed as calcium carbonate.

Hardness in water poses problem of scale formation in boilers. It is also objectionable for laundry and domestic purposes since it consumes a larger amount of soap.

Hardness may range from zero to hundreds of milligrams per liter in terms of calcium carbonate, depending on the source of water.

### EDTA Titrimetric Method

***(i) Principle***

At a pH $10 \pm 1$ EDTA forms a soluble chelated complex with calcium and magnesium. Calcium and complexed magnesium ions also forms a less stable complex with Eriochrome Black T indicator. Thus, when EDTA is added to a flask, metal-indicator complex breaks to form a more stable metal-EDTA complex and when all the metal is complexed to EDTA indicator shows a sharp change in colour from wine red to blue. Magnesium ion must be present to yield a satisfactory and point in the titration. A small amount of complexometrically neutral magnesium salt of EDTA is therefore added to the buffer. Thus, this step automatically introduces sufficient magnesium and at the same time obviates a black correction.

### *(ii) Interference*

Some metal ions interfere in the end point detection which can be overcome by addition of inhibitors. Usually this interference is unlikely in surface waters.

### *(iii) Reagents*

*Buffer solution*

(1) Dissolve 16.9 g Ammonium chloride, in 14.3 ml conc. Ammonium hydroxide, add 1.25 g of magnesium salt of EDTA and dilute to 250 ml with distilled water.

(2) If the magnesium salt of EDTA is not available, dissolve 1.179 g disodium salt of ethylenealimina tetra acetic acid dihydrate (analytical reagent grade) and 780 mg $MgSO_4$. $7H_2O$ or 644 mg $MgCl_2.6H_2O$ in 50 ml distilled water. Add this solution to 16.9 g $NH_4Cl$ and 143 ml conc. $NH_4OH$ with mixing and dilute to 250 ml with distilled water.

Store the solution 1 and 2 in a plastic container. Do not store more than a month's supply in a frequently opened container.

### *(iv) Indicator*

Mix 0.5 g of Eriochrome Black T indicator with 100 g of solid sodium chloride. Grind together in a morter to obtain a homogenious mixture of 40 to 50 mash size. Use about 0.2–0.4 g of the ground mixture for each titration.

*Standard EDTA titrant*

0.01 N dissolve 3.72 mg analar reagent grade disodium salt EDTA in distilled water, and dilute to 1 litre, standarize against standard calcium. This solution is about 1.00 mg $CaCO_3$/ml.

*Standard Calcium solution*

0.01 M weigh 1.000 g anhydrous calcium carbonate into a 500 ml conical flask. Place a funnel on the neck of the flask and add slowly 1 N HCl until all the $CaCO_3$ has dissolved. Add 200 ml distilled water and boil for a few minutes to expel out all $CO_2$.

Cool, add a few drops of methyl red indicator and adjust to the intermediate orange colour by adding 3N NaOH or 1N HCl, as required. Transfer quantitatively to a one litre volumetric flask and make-up the volume to the mark. This standard solution is equilavant to 1.00 mg $CaCO_3$/ml.

***(v) Procedure*** **:** Take 100 ml sample or smaller appropriate portion diluted to 100 ml so that hardness is in range 10 to 20 mg, in a porcelain dish or conical flask. Add 3 ml buffer solution followed by about 0.2-0.4 g of solid indicator. Titrate immediately but slowly with continuous stirring, until the last reddish tinge disappears and blue colour is observed. Reagent blank may be used for comparison. Reagent blank is titrated in a similar way as for the sample.

If the hardness is too low, larger volume may be used and also increase the other reagents and indicator proportionately. Blank may be carried out under identical conditions.

Standarize EDTA titrant with 25 ml calcium standard in a similar way as described for the sample.

***(vi) Calculations***

$$\text{Hardness as mg/l } CaCO_3 = \frac{(V_1 - V_2) \times 1000}{\text{ml sample}} \times \frac{25}{V_3}$$

Where,

$V_1$ is the volume of EDTA required for sample.

$V_2$ is the volume of EDTA required for reagent blank.

$V_3$ is the volume of EDTA required for 25 ml standard Calcium.

***(vii) Expression of Results***

Hardness is expressed as mg/l $CaCO_3$. Round off the results to nearest whole number.

***(viii) Precision and Accuracy***

The precision and accuracy depends on hardness concentration. When the concentration is more than 25 mg/l, precision and accuracy of about ± 2 to 4% can be achieved.

***(ix) Precautions***

1. Carry out the titration within 5 minutes to minimize the tendency toward $CaCO_3$ precipitation.

2. The final volume of sample and blank should be kept constant. Usually 100 ml is better.
3. Use deionized or double distilled water for dilutions, blank estimation, etc.
4. Use reagent blank for colour comparison.
5. Add same quantity of indicator for sample and blank.
6. Adjust volume of sample such that hardness is in range 5 to 20 mg.

## 5. CHLORIDE

Chloride is one of the major inorganic anions in water and waste water. A high concentration occurs in waters from chloridecontaining geological formations. Otherwise, a high chloride content may indicate pollution by sewage or some industrial wastes or an instruction of sea water or other saline water.

The salty taste produced by chloride depends on the chemical composition of the water. A concentration of 250 mg/l may be detectable in some water containing sodium ions. On the other hand, the typical salty taste may be absent in waters containing 1000 mg/l of chloride when the predominant cations are calcium and magnesium.

Although chlorides are not harmful, concentration beyond 250 mg/l impart a peculiar taste to water rendering it unacceptable from aesthetic point of view for drinking purpose. A high chloride concentration harms metellic pipes and structures, as well as agricultural plants.

Two methods, both titrimetric argentrometric and mercurimetric method are described. Wherever colour of the sample interference with the chromate and point in argentrometric method, the other method may be used.

### (A) Argenometric Method

#### *(i) Principle*

Chloride is determined in a neutral or slightly alkaline solution by titration with standard silver nitrate, using potassium chromate as indicator. Silver chloride is precipitated quantitatively before red silver chromate is formed.

### *(ii) Interferences*

Bromide, iodide and cyanide are measured as equivalent to chloride ion. Sulfide, thiosulphate, cyanide thiocynate and sulfites interfere. They may be oxidised to noninterfering substances as described below :

Take a known volume of sample, add 25 ml of hydrogen peroxide 3% and boil for 15 minutes. Add further 10 ml of hydrogen peroxide and boil for 5 minutes. Test byspotting out for the absence of thiocynate with ferric chloride. Formation of yellowish orange spot indicates presence of thiocyanate. Repeat the heating with hydrogen peroxide till it becomes free from thiocyanate. Orthophosphate in excess of 25 mg/l interferes by precipitation as silver phosphate. Iron in excess of 10 mg/l interferes by making the end point.

### *(iii) Reagents*

*Silver nitrate standard solution, 0.0282N*

Dissolve 4.791 silver nitrate in distilled water and dilute it to 1000 ml. Store in brown glass bottle.

$1.0 = 1.0$ mg $Cl^-$

Standarization titration must be carried out without any dilution.

*Potassium chromate indicator solution*

Dissolve 50 g $K_2CrO_4$ in a little distilled water. Add silver nitrate solution dropwise until definite red precipitation starts. Let it stand for 12 hrs, filter, and dilute to 1 litre with distilled water.

Use chloridefree ie redistilled or deionized distilled water for the preparation of all above reagents. Some water should be used for the sample dilutions also (wherever required). Avoid dilutions as far as possible or apply blank correction.

### *(iv) Procedure*

Take 100 ml sample or a suitable portion diluted to 100 ml into a porcelain dish, or conical flask or beaker on a white surface.

Titrate samples in the pH range 7 to 10 directly. Adjust samples not in this range with $H_2SO_4$ or NaOH solution. Add 1 ml potassium chromate solution and titrate with silver nitrate solution with constant stirring until only the slightest reddish colouration persists. Be consistent in and point recognition. It is suggested to use reagent and blank as a comparator. If the volume of the titrant exceeds 25 ml then it is better to repeat the method with smaller portion of the sample.

Standarize the silver nitrate titrant with standards sodium chloride. Carry out the blank determination to account for the chloride present in any of the reagents and for the aclubility of silver chromate. A blank of 0.2 to 0.4 is usual for the method.

### *(v) Calculation*

$$\text{Chlorides Cl}^- = \frac{(V_1 - V_2) \times 1000}{\text{Volume of sample (ml)}} \times \frac{25}{3}$$

Where,

$V_1$ = Volume of silver nitrate required for sample

$V_2$ = Volume of silver nitrate required for reagent blank.

$V_3$ = Volume of silver nitrate required for 25 ml of standard sodium chloride.

### *(vi) Expression of results*

Results are expressed in $mgl^{-1}Cl^-$

The results should be rounded off as follows :

Concentration results mg $l^{-1}$ less than 20 above 20

Record to nearest $mgl^{-1}$ 0-1 to nearest whole number

### *(vii) Precision and Accuracy*

The precision and accuracy of about 2% to 4% is expected from a good analyst, except in the samples were chloride content is less than 50 mg/l.

### *(viii) Precautions*

(1) Use white opaque background for titration, it is suggested to carry out titrations in white porcelain dish.

(2) The final volume of titrant exceeds 25 ml, it is suggested to take a smaller quantity of the sample.

(3) Also the volume of titrant should not be too small in such cases volume of sample may be increased. It is better to adjust the volume of sample such that titrant value is in range 5-20 ml.

(4) Sample is coloured which may interfere with the chromate end point in such cases mercurimetric method should be used.

(5) Be consistent in endpoint recognition. Reagent blank may be used for comparison.

(6) Use deionized distilled water for the preparation of solution as well as any dilutions.

## (B) Mercuric Nitrate Method

### *(i) Principle*

Chloride ions from a highly stable complex with mercuric ions, thus they may be titrated with a standard mercuric nitrate. The end point is detected by diphenylcarbazone indicator in the pH range 2.9 to 3.4 which forms a blue violet complex with an excess of mercuric ions.

### *(ii) Interference*

Iodide, bromide, cyanide and thiocyanide are measured as equivalents of chlorides. Sulfide, thiosulphate, sulfite, cyanide and thisulphate interfere seriously with the determination. They may be oxidised to noninterfering substances as described in argentrometric method.

If the sample is highly coloured or turbid to allow the end point to be readily seen this interference may be reduced by the aluminium hydroxide suspension treatment.

Add 3 ml aluminium hydroxide suspension to the measured quantity of sample. Stir thoroughly, lot is settle for few minutes and filter. Wash the precipitate with 10-15 ml distilled water. Also collect washing with the filtrate and continue as described under procedure.

Chromate, ferric, zinc, copper, mangenous and sulfite ions interfere in concentrations above 10 mg/l, aluminium, lead and nickel in concentrations above 100 mg/l. In such cases these titrimetric methods are not suitable.

Phosphate do not interfere.

### *(iii) Reagents*

*Mercuric nitrate solution 0.0282N*

Dissolve 5.048 g mercuric nitrate, $Hg(NO_3)_2\ H_2O$, in 50 ml water containing 0.5 ml*. This solution is approximately 1.0 ml =

1.0 mg $Cl^{-1}$. Standarize against 25 ml of standard sodium chloride without any dilution. Carry out reagent blank with 25 ml of deionized distilled water.

*Conc. nitric acid dilute to 1 litre and filter if necessary.

*Sodium Chloride Standard Solution 0.282N*

Dissolve 1.648 g dried sodium chloride in distilled water and dilute to 1 it.

1.0 ml = 1.0 mg $Cl^-$

***(iv) Indicator***

Dissolve 0.5 g diphenylcarbazone and 0.05 bromophenol blue in 100 ml alcohol. Store in brown bottle.

*Nitric acid, approximately 0.05 N*

Dilute 3.2 ml conc. nitric acid (density 1.42 g/c.c.) to 1 litre.

*Sodium hydroxide, approximately 0.05 N*

2 g sodium hydroxide in distilled water and dilute to 1 litre.

**(C) Aluminium hydroxide suspension**

Dissolve 100 g potassium aluminium sulphate $K_2SO_4$. $Al_2SO_4)_3$. 24 $H_2O$, in about 1 litre of water. Precipitate the aluminium hydroxide by slowly adding ammonium hydroxide until it is in slight excess, with stirring. Allow the precipitate to settle, wash by decantation and check that the final decantate is free from chloride. Dilute the suspension of aluminium hydroxide to about 1 litre.

***(i) Procedure***

Take 100 ml or smaller portion sothat the chloride content is less than 20 mg, in a porcelain dish. Add 510 drops of indicator. If a blue, violet or red colour develops, add 0.05 N nitric acid drop by drop until it becomes yellow and then add exactly 1 ml in excess. Titrate the solution with standard mercuric nitrate to the point where the first tinge of blueviolet colour appears and which does not disappear on stirring. Prior warning that the end point near is given by the appearance of orange colour.

Determine the reagent blank by titrating 100 ml of distilled water by the same procedure.

### *(ii) Calculation*

$$\text{Chlorides as } Cl^- \ \frac{(V_1 - V_2) \times 1000}{\text{Volume of sample (ml)}} \times \frac{25}{V_3} \text{ mg/l}$$

Where,

$V_1$ = Volume of mercuric nitrate required for sample

$V_2$ = Volume of mercuric nitrate required for reagent blank.

$V_3$ = Volume of mercuric nitrate required for 25 ml of a standard sodium chloride.

### *(iii) Expression of results*

Record the results as mg/l $Cl^-$

The result should be rounded off as follows :

| Concentration range (mg $l^{-1}$) 1-20 | 20 onwards |
|---|---|
| Record to nearest (mg $l^-$) 0.1 | To nearest (whole number) |

### *(iv) Precision and Accuracy*

The precision and accuracy of about 2 to 4% is expected from a good analyst. This also depends on quantity of chloride present, and on the pretreatment necessary.

### *(v) Precautions*

1. Titration should be carried out on a opaque white base for which white porcelain dish may be used.
2. To final volume should be kept constant say 100 ml.
3. The volume of the sample should be selected such that it should not contain more than 20 mg of chloride. On the other hand, volume of sample may be increased if the chloride content is too small.
4. Deionized (Chloride free) distilled water should be used for the preparation of solutions as well as for dilution, etc.

## 6. ALKALINITY

***General :*** The alkalinity of a water is its quantitative capacity to neutralize a strong acid to a designated pH. Since the alkalinity of many surface water is primarily a function of carbonate, bicarbonate and hydroxide ions, the alkalinity is taken as an indication of the concentration of these constituents. The measured values may also include contributions from borates, phosphates or silikates, if these are present.

The alkalinity of many neutral and treated waters is due only to bicarbonates of calcium and magnesium. The pH of these waters does not exceed 8.3. Their total alkalinity is practically identical with their carbonate hardness and such samples do not have phenolphthelein alkalinity.

Water having a pH above 8.3 contain besides bicarbonates, normal carbonates and possibly hydroxides also. The alkalinity fraction equivalent to the amount of acid required to lower the pH of the sample to 8.3 is called phenolphthalein alkalinity. This fraction is contributed by half of the carbonate and by the hydroxide, if present.

The amount of acid required to lower down the pH to 4.5 determines the total alkalinity or methyl orange alkalinity.

Alkalinity values provide guidance in applying proper losses of chemicals in water and waste water treatment process, particularly in coagulation, softening and operational control of anaerobic digestion. The alkalinity in excess of alkaline earth concentrations is significant in determining the suitability of a water in irrigation.

## METHODS

The alkalinity is determined by titration of the sample with a standard solution of a strong mineral acid. Titration is carried out either using indicator or a calibrated pH-meter.

The simple and rapid visual method using an indicator is satisfactory for control and routine applications. Electrometric method is used for accurate determinations. Electrometric method is used for accurate determinations. It is also free from interference such as colour, turbidity or suspended matters. Low alkalinities (below 20 mg/l $CaCO_3$) are also best determined by electrometric titration. Thus, for the natural or surface waters were low alkalinity is expected electrometric method is preferred over the visual indicator method.

Wherever possible, the titration should be carried out at the point of sampling. Otherwise, the sampling bottle should be completely filled and the alkalinity is determined within 24 hours.

### *Interferences*

Colour, turbidity and suspended matter may interfere with the visual titration method by masking the colour change of the indicator. Turbidity and suspended matter can be eliminated by simple filtration.

Electrometric method is free from all above interferences.

## (A) Electrometric Titration Method

### *(i) Apparatus*

Any pH meter utilizing a glass electrode may be used. It must be standardized according to the directions supplied by the manufacturer. It is convenient to use a burrette with a bent jet, sothat the burrete can be used in an upright position.

### *(ii) Reagents*

*Carbon dioxide free distilled water*

It must be used for the preparation of all standard solutions. If the distilled water has a pH lower than 6.0, it should be freshly boiled for 15 minutes and cooled to room temperature.

*Sodium carbonate solution 0.05 N*

Dilute 3.1 ml conc. sulfuric acid or (10 ml conc. hydrochloric acid) to 1 litre. Standarize against 40.0 ml 0.05N $Na_2CO_3$ solution, with about 60 ml water, in a 250 ml beaker by titrating electrometrically to the pH inflection point. This can also be standardized using methyl orange indicator.

*Sulfuric acid (or hydrochloric acid) 0.02 N*

Dilute 200 ml of 0.1 N standard acid to 1 litre with carbon dioxide free distilled water. Standarize using 10 ml of 0.05 N $Na_2CO_3$ by electrometric or visual titration method. Calculate the normality from equation given below :

$$\text{Normality of acid, N} = \frac{(X) \times (Y)}{53{,}00 \times (Z)}$$

Where,

X is g $Na_2CO_3$ dissolved in 1 litre.

Y is ml $Na_2CO_3$ taken for titration.

Z is ml standard acid consumed in titration.

When concentration of sulfuric acid is adjusted exactly to 0.1000 and 0.0200 N.

1 ml of 0.1000 N solution = 5.00 mg $CaCO_3$

1 ml of 0.0200 N solution = 1.00 mg $CaCO_3$

***(iii) Procedure***

Take 100 ml sample in a 250 ml beaker, place it on a magnetic stirrer. Dip the electrodes into the solution and adjust the tip of the burette on the surface of the sample. Titrate the standard acid to pH 8.3. Record the volume required, this corresponds to the phenolphthalein alkalinity. Continue the titration to a pH of 4.5. The total titrations value represents the total alkalinity.

**(B) Visual Titration Method**

***(i) Reagents***

*Carbon dioxide-free distilled water*

Same as described for the electrometric method.

*Sodium carbonate 0.05 N*

As described under the previous method.

*Phenolphthalein indicator*

Dissolve 0.5 g of phenolphthalein in 50 ml of 95% ethanol and add 50 ml of distilled water. Add to it dilute (e.g. 0.01 N or 0.05 N) carbon dioxide free solution of sodium hydroxide slowly drop by drop till a faint pink colour persists.

*Methyl Orange Indicator*

Dissolve 0.5 g of methyl orange in 100 ml water.

***(ii) Procedure***

Add 3-4 drops of phenolphthalein indicator to 100 ml of water, in a porcelain dish or a conical flask over a white surface. If no colour is produced, the phenolphthalein alkalinity is zero. If the sample turns pink, titrate with standard acid till pink colour just disappears. In both cases continue the determination using the sample to which phenolphthalein has been added.

Add a few drops of methyl orange indicator, if orange colour is observed without adding any acid, the total alkalinity is zero. If the sample turns yellow, titrate with standard acid until the first appearance of orange colour is noticed.

***(iii) Calculation***

$$\text{Phenolphthalein alkalinity (pH 8.3) mg/l as } CaCO_3 = \frac{A \times N \times 50{,}000}{\text{ml of sample}}$$

$$\text{Total Alkalinity (pH 4.5) mg/l as } CaCO_3 = \frac{B \times N \times 50{,}000}{\text{ml of sample}}$$

Where,

A is the ml of standard acid required to reach the Phenolphthalein end point or pH 8.3

B is the ml of standard acid required to reach the methyl orange end point or pH 4.5

N is the normality of acid used for the titration

***(iv) Expression of results***

Results are expressed as mg/l of $CaCO_3$. For conversion purposes it may be recalled that 50 mg/l as $CaCO_3$ are equivalent to 1 m Eq/l.

***(v) Precision and Accuracy***

No general statement can be made about precision because of the great variation in sample characteristics.

For total alkalinity higher than 50 mg/l, precision and accuracy of electrometric titration are estimated around 1% while for the lower concentrations it may be around 2% to 3%.

The precision and accuracy of visual titration is estimated at 2% for the alkalinity above 50 mg/l as $CaCO_3$ and 2%-10% for alkalinity between 50 mg/l and 5 mg/l and 5 mg/l as $CaCO_3$.

***(vi) Precautions***

(1) Do not dilute, concentrate or alter the sample in any way. As far as possible also avoid filtration.

(2) Use only carbon dioxide free distilled water for all purposes such as preparation of reagents, etc.

(3) If possible titrations should be carried out on the spot. If not, the sampling bottle should be filled completely and the

alkalinity is determined in shortest possible time, in any case not later than 24 hours.

(4) Use magnetic stirrer for stirring purposes.

(5) Adjust the volume of sample and concentration of titrant such that the volume of titrant required is appropriate i.e. neither too low nor too high.

(6) While preparing 0.1 N $H_2SO_4$, add conc. $H_2SO_4$ to water (not water to $H_2SO_4$) cool then make up the volume.

## 7. RESIDUE

A highly mineralized water is less acceptable then a water of little or moderate mineral content for drinking, household and special industrial purposes. A good indication of measuring the quantity of minerals present in water or waste water can be determined by finding out the material left after evaporation.

The term "Residue" applies to the material remaining in the dish after evaporation and drying of a water sample in an oven at a definite temperature. The term "Suspended solids" and "Dissolved Solids" correspond to nonfilterable and filterable residue, respectively. "Fixed Solids" and "Volatile Solids" correspond to the remainder after ignition and loss due to ignition at a definite temperature, respectively.

### (A) Total Residue

*Principle* : A well-mixed sample is evaporated and dried in a weighted china dish in an oven at 105 + 1□C. The increase in weight represents the total residue.

#### *(i) Apparatus*

(a) Evaporating dishes of 150-200 ml capacity made of porcelain, alternatively, beakers can be used.

(b) Drying oven at 105 + 1□C.

(c) Desiccator, provided with a desiccant.

(d) Analytical balance capable of weighing up to 0.1 mg.

(e) Muffle furnace at 550 to 560□C.

#### *(ii) Procedure*

(1) Heat the clean evaporating dish at 550 ± 5□C for 1 hr in a muffle furnace.

(2) Cool, desiccate, weight and store the dish in a desicator until it is used. On repetition of this process dish should show a constant weight.

(3) Transfer a known volume of wellmixed sample to the preweighed dish and evaporate to dryness in the oven at 105 + 1□C. Choose a sample volume in such a way that it will yield a residue between 25 and 250 mg and preferably between 100 and 250 mg (usually, 50 to 100 ml for wastewater and 100 ml to 200 ml for clean waters is suitable).

(4) After complete evaporation dry it atleast for 1 hr. in the oven.

(5) Cool the dish in a desiccator and weigh it.

(6) Repeat the cycle of drying, cooling and weighing till the weight of the dish becomes constant.

***(iii) Calculation***

$$\text{Total residue (mg/l)} = \frac{(X - Y) \times 10^6}{V}$$

Where,

X = Weight of dried sample + dish in g

Y = Weight of empty dish/in g

V = Volume of sample in ml.

***(iv) Expression of results***

Report the results as "total residue" or "total solids" in mg/l. Round off the results to the nearest whole number.

***(v) Precision and Accuracy***

The precision is about ± 4 mg or ± 5% for raw waters, the results are considered statistically unreliable due to sampling errors.

**(B) Fixed Solids and Volatile Solids**

***(i) Principle*** **:** The residue produced on evaporation (at 105 ± 1□C in oven) is ignited in a muffle furnace at 550-560□C. The remainder and the loss on ignition represent fixed and volatile solids, respectively.

***(ii) Apparatus*** — Same as per "Total Residue"

***(iii) Procedure***

(1) After completion of the test for total residue, ignite the dish with the residue in a muffle furnace at 550-560°C for about 30 minutes (30 minutes after muffle furnace has attained the temperature of 550-560°C).

(2) Allow the dish to cool partially in air most of the heat has dissipated or switch off the muffle furnace and let the temperature fall.

(3) Transfer the dish to a desiccator, cool and weigh it as soon it has cooled completely.

(4) Repeat the operation of ignition, cooling and weighing until a constant weight is obtained (i.e. till the difference between two successive series of operation is not more than 0.5 mg or 1% of the remainder solids).

***(iv) Calculation***

$$\text{Volatile Residue, mg/l} = \frac{(X - Y) \times 10^6}{V}$$

$$\text{Fixed Residue, mg/l} = \frac{(Y - Z) \times 10^6}{V}$$

Where,

X = Weight of sample + dish, before ignition but after evaporation in g.

Y = Weight of sample + dish after ignition in g.

Z = Weight of dish in g.

V = Volume of sample in ml.

***(v) Expression of Results***

Report the results as "fixed residue" or "fixed solids" and "Volatile residue" or "volatile solids"in mg/l. Round off the results to the nearest whole number.

***(vi) Precision and Accuracy*** — Same as per "Total residue".

**(C) Dissolved Solids**

***(i) Principle :*** Dissolved solids or filterable residue, are those solids capable of passing though a standard glass fibre filter paper. (In waters containing no suspended solids, the dissolved solids are equal to the total solids). The filtrate from the nonfilterable residue or suspended solids can be used to determine the filterable residue.

Fixed dissolved solids and "Volatile dissolved solids" correspond to the filterable solids remainder after ignition and loss by ignition, respectively.

***(ii) Apparatus***

(1) Glass-fibre filter paper, Whatman GF/C grade or equivalent.

(2) Millipore filtration assembly with suction flask of 500 ml or 1000 ml capacity.

(3) Porcalein dish or beaker, 150-200 ml capacity.

(4) Drying oven, 180°C ± 2°C or 105°C ± 1°C.

(5) Desiccator

(6) Analytical balance capable of weighing upto 0.1 mg.

***(iii) Procedure***

(1) Preparation of evaporating dish : Heat the clean dish to 550-560°C for about one hour in a muffle furnace. Cool, partially in air then in desiccator. Weigh immediately before use.

(2) Preparation of glassfibre disc : Place the disc on the membrane filter holder. Wash it with three successive 20 ml volume of distilled water. Discard all these washings.

(3) Under vacuum filter 100 ml or more of well mixed same through glass fibre filter.

(4) Transfer 100 ml (or a larger volume if total filterable residue is low) filtrate to a preweighed evaporating dish.

(5) (a) Dissolved Solids at 105°C : Evaporate and dry the sample in a dry oven at 105 ± 1°C (Dry atleast for 1 hr after complete evaporation). Cool in a desiccator and weigh.

Repeat the drying cycle until a constant weight is obtained. (i.e. ± 0.5 mg or ± 1% of the evaporated sample).

(b) Dissolved Solids at 180°C : The above procedure to be carried out at 180 ± 2°C instead of 105 ± 1°C.

**(iv) Calculation**

$$\text{Total dissolved solids, mg/l} = \frac{(X - Y) \times 10^6}{V}$$

Where,

X = Weight of dried filtered sample + dish in g

Y = Weight of empty dish in g.

V = Volume of sample in ml.

**(v) Expression of results**

Express the results as "Total dissolved solids" dried at – ▯C, mg/l.

**(D) Volatile and Fixed Dissolved solids**

***(i) Procedure*** : 1. Determine the proportion of matter volatile on ignition by igniting the dried residue (after during dissolved solids) in a muffle furnace at 550▯C for about an hour.

2. Cool in a desiccator and weigh it.

***(ii) Calculations***

$$\text{Volatile dissolved solids, mg/l} = \frac{(X - Y) \times 10^6}{V}$$

Where,

X = Weight of dried residue + dish, g (before ignition)

Y = Weight of ignited residue + dishg (after ignition)

V = Volume of filtrate taken ml.

$$\text{Fixed dissolved solids, mg/l} = \frac{(Y - Z) \times 10^6}{V}$$

Where,

Y = Weight of ignited residue + dish, g

Z = Weight of dish, g

V = Volume of filtrate taken ml.

***(iii) Expression of results***

Express results as "Volatile dissolved solids" and "Fixed dissolved solids" as mg/l. Round off the results to the nearest whole number.

**(E) Suspended Solids**

***(i) Total suspended solids*** **:** Suspended solids are those solids which are retained on a standard glass fibre filter disc.

***(ii) Samples handling and preservation*** :

Non-homogenous particulate matter such as leaves, sticks, fish and lumps of focal matter should be excluded from the sample.

Preservation of the sample is not practical, analysis about be carried as soon as possible.

***(iii) Apparatus***

(a) Glassfibre filter disc, Whatman GF/C grade or equivalent, 70 mm

(b) Millipore filtration assembly with suction flask

(c) Drying oven at 105 ± 1°C

(d) Desiccator

(e) Analytical balance capable of weighing to 0.1 mg

***(iv) Procedure***

Preparation of glass fibre filter disc :

(1) After numbering the filter disc dry it in an oven at 105 ± 1°C for about an hour. Cool in a desiccator and weigh it.

(2) After weighing place it on the filter holder. Apply vacuum and wash it with about 20 ml of distilled water.

While the pump is on transfer 100 ml or an appropriate of volume of the wellmixed sample so that it contains about 25 to 200 mg ms of suspended solids, to the filtration of the milipore assembly. If the solids are very high add small volume in number of installments.

After complete filtration of the sample add about 20 ml distilled water.

Transfer the filter paper carefully to the petridish when it is complete dry. Dry at least one hour at 105 ± 1°C in the oven. Cool in a desiccator and weigh.

Repeat the drying cycle until a constant weight is obtained.

***(v) Calculation***

$$\text{Total suspended solids, mg/l} = \frac{(X - Y) \times 10^6}{V}$$

Where,

X = weight of filter paper + residue, g

Y = weight of filter paper, g

V = volume of sample, ml

### (vi) Expression of results

Express the results as total suspended solids in mg/l. Round off to the nearest whole number.

## (F) Volatile and Fixed suspended solids

### (i) Procedure

(1) After determining the total suspended solids the filter disc containing the suspended matter is ignited in the muffle furnace at 550▯C for about an hour.

(2) Cool in a desiccator and weigh.

### (ii) Calculation

$$\text{Volatile suspended solids, mg/l} = \frac{(X - Y) \times 10^6}{V}$$

Where,

X = weight of dried residue + filter disc, g (before ignition)

Y = weight of ignited residue + filter disc, g (after ignition)

V = volume of sample filtered, ml.

$$\text{Fixed suspended solids, mg/l} = \frac{(Y - Z) \times 10^6}{V}$$

Where,

Y = weight of ignited residue + filter disc, g

Z = weight of filter disc, g

V = volume of sample filtered, ml

### (iii) Expression of results

Express the results as volatile suspended solids and fixed suspended solids as mg/l. Round off to value the nearest whole number.

### (iv) General Precautions

(1) Always set the balance to zero before use.

(2) Before weighing ensure that dish or filter disc is cooled to the room temperature.

(3) At all stages prevent dish or filter disc from dust, etc.

(4) Greater the number of replicates more is the accuracy. Keep at least 3 replicates.

(5) Mark the number on the dish or filter disc before altering the test (before drying).

## 8. DISSOLVED OXYGEN (DO)

***(i) General*** **:** Dissolved oxygen (DO) in natural and waste water depends on the physical, chemical and biological activity in a water body. The analysis for DO is important in water pollution control and waste treatment process control. DO as a parameter used :

(a) To asses quality of surfacewater and to keep a check on stream pollution.

(b) It is the basis for BOD which is an important parameter to evaluate pollution potential of wastes.

(c) It is responsible for corrosivity, photosynthetic activity and septicity.

(d) DO is necessary for all aerobic biological wastewater treatment processes.

The iodometric test is precise and reliable titrimetric procedure for DO analysis. The test is based on the addition of divalent manganese solution, followed by strong alkali, to the sample in a BOD glass stoppared bottle. DO present, rapidly oxidizes an equivalent amount of the dispersed divalent manganous hydroxide of higher oxidation state (brownish-orange in colour). This brownish-orange colour indicates presence of oxygen. In the presence of iodide and on subsequent acidification, higher manganese hydroxide revert to the divalent state, and liberate iodine equivalent to the original dissolved oxygen content of the sample. The iodine liberated is titrated with a standard solution of sodium thisulphate using starch as an indicator.

***(ii) Interference***

It is caused by certain oxidizing agents which liberate iodine from iodides. (Such as ozone, chlorine, ferric compounds, manganese in tri or higher oxidation states chromate, nitrate, persulphate, peroxides, etc. by certain reducing agents which may reduce iodine to iodide (Such as Ferrous compounds, thiosulphate, readily oxidizable organic matter, nitrite etc.). Some organic compounds hinder the setting of the precipitate and colour of some may interfere in the starch end point. Modifications are included for minimize the effect of interfering materials.

### *(iii) Apparatus*

The bottle used should be of good quality having 300 ml capacity with narrow neck and well fitting ground glass stopper. It is convenient to have number on bottles and their stoppers.

### *(iv) Reagents*

(a) *Mangenous sulphate solution* : Dissolve 500 g $MnSO_4$ in distilled water. Filter and dilute to 1 litre. This mangenous sulphate solution should not give a colour with starch when added to an acidified solution of potassium iodide.

(b) *Alkali-iodideazide reagent* : Dissolve 500 g NaOH (or 700 g KOH), and 135 g NaI, (or 150 g KI) in distilled water and dilute to 1 litre. Add 10 g sodium azide, $NaN_3$ dissolved in 40 ml distilled water. Potassium and sodium salts may be used interchangeably. This reagent should not give colour with starch solution when diluted and acidified.

(c) *Sulfuric acid* : $H_2SO_4$ conc. one milliliter of this is equivalent to about 3 ml of alkaliiodideazide reagent.

(d) *Starch indicator* : Make a smooth paste of 5 g soluble starch in cold water and pour this into 1 litre boiling water with constant stirring. Boil for one minute and allow to cool before use. Use clear supernatant. Preserve with 1.25 g salicylic acid /l or by adding a few drops of toluene.

(e) *Sodium thiosulphate stock solution 0.125 N* : Dissolve 31.5 g sodium thiosulphate pentahydrate, $Na_2S_2O_3$, $5H_2O$ in 1 litre copper free boiled and cooled distilled water. Add 5ml chloroform or 1g NaOH as preservative.

(f) *Standard sodium thiosulphate titrant 0.125 N* : Dilute the stock sodium thiosulphate solution to obtain 0.0125 (100 ml of stock 0.125 M $Na_2S_2O_3$ diluted to 1000 ml with distilled water gives 0.0125 N $Na_2S_2O_3$) standarize this diluted solution daily.

### Standarization

Pipette 10 ml 0.012 5N Potassium iodate into a conical flask containing about 100 ml distilled water. Add 2 ml of conc. $H_2SO_4$ followed by about 2 g of potassium iodide solid. Titrate immediately against sodium thiosulphate 10/25N using starch as indicator. Apply concentration correction in the calculation.

*Potassium iodate, 0.0125 N*

Dissolve 0.146 potassium iodate previously dried at about 120□C, in distilled water and dilute to 1 litre. The solution is stable for a long period if stored in glassstoppared bottles.

***(v) Procedure***

(a) Add 2 ml mangenous sulphate followed by 2 ml alkali-iodide-azide reagent, dipping the pipette little below the surface.

(b) Stopper carefully to exclude air bubbles and thoroughly mix the contents. To attain this invert the bottle at least 15 times.

(c) Allow the precipitate to settle to the lower one third of the bottle, repeat the mixing and allow the precipitate to settle completely leaving a clear supernatant liquid.

(d) Add 2 ml conc. $H_2SO_4$ immediately after removing the stopper. Restopper, and mix by gentle inversion until all the precipitate dissolves, if it does not, allow to stand for a few minutes and repeat the mixing.

(e) Distribute the iodine uniformly before taking a part of it titration.

(f) Take 100 ml of sample for titration (1.3 ml extra to account for 4 ml or the reagents, mangenous sulphate and alkali-iodide-azide, added to 300 ml of sample) in a conical flask and immediately titrate the liberated iodine with the standard thiosulphate solution to a pale yellow straw colour. Add 2 ml starch solution and continue the titration till the first disappearance of the blue colour. Disregard subsequent recolouration due to the catalytic effect of nitrite of ferric salts.

It must be remembered that iodine is volatile and therefore the titration must be carried out as expediously as possible and with the minimum of exposure to the air.

RIDEAL-STEWART MODIFICATION FOR ELIMINATION OF INTERFERENCE BY FERROUS IRON

This modification is used when the sample contains ferrous iron. Add 0.7 ml conc. $H_2SO_4$ followed by 1 ml potassium permanganate (0.63%) and 1 ml potassium floride (40%) to the sample bottle, stopper and mix by inversion. The amount of permanganate added should be just sufficient to obtain a violet tinge that persists for 5 minutes. Remove the excess permanganate by adding 0.5 to 1.0 ml potassium oxalate (2%). Mix well and let stand in the dark to fecilitate the reaction. Excess oxalate causes low results and hence add only an amount of oxalate that completely decolourizes the potassium permanganate.

After applying this modification usual method should be followed.

***(vi) Calculation***

$$\text{Dissolved oxygen Content mg/l} = \frac{\text{Volume of thiosulphate required for 101.3 ml sample x 10}}{\text{Volume of thiosulphate required for 10 ml of 0.0125 N Potassium iodate}}$$

***(vii) Expression of results***

Express the dissolved oxygen results in mg/l. Report the values to the nearest of first decimal i.e. 0.1 mg/l.

***(viii) Precision and Accuracy***

The DO can be measured with a precision of about 50 μg/l or 0.05 mg/l in surface and slightly polluted waters, expressed as standard deviation. In the presence of appreciable interference this may be as high as 100 μg/l or 0.1 mg/l.

***(ix) Precautions***

(a) Do not use a cap or synthetic detergents for washing the DO bottles. Normally bottles are kept clean by the acidic iodine solution of the winkler procedure and require no further treatment except thorough rinsing with the tap water.

New bottles should be cleaner with 5N sulfuric or hydrochloric acid and then rinsed thoroughly with the tap water.

(b) Just before taking the sample rinse sample bottle with the sample.

(c) Bottle should be completely filled with the sample and remove bubbles entrapped into the bottle while taking sample by tapping the neck of the bottle with the stopper.

(d) Fix the DO on the site by adding manganous sulphate and alkaliiodideazide solution.

(e) The titrations should be carried out as expediously as possible and with minimum exposure to the air, so as to avoid any losses of iodine during titration.

(f) Use BOD samples discharger to avoid inconvenience caused in measuring 101.3 ml. This will save time as well as maintain uniformity in the comparative analysis.

(g) Avoid errors due to carelessness in collecting the samples, prolonging the completion of the test or due to any interferences.

## 9. BIOCHEMICAL OXYGEN DEMAND (BOD)

### *(i) General*

Biochemical oxygen demand (BOD) is defined as the amount of oxygen required by micro organisms to stablize biological decomposable organic matter in a waste under aerobic conditions. The biochemical oxygen demand is an approximately measure of the amount of biochemically degradable organic matter present in the sample.

The BOD test is widely used to determine (1) the degree of pollution in lakes and streams at any time and their self purification capacity, (2) the pollutional load of waste waters and (3) efficiency of waste treatment plants.

Since the test is based on mainly bioassay procedure measuring amount of oxygen consumed by bacteria while stabilizing organic matter under aerobic conditions, it is necessary to provide standard conditions of nutrient supply, pH, absence of substances inhibiting microbial growth and temperature. Because of the low solubility of oxygen in water, the sample is diluted proportionately depending upon the expected BOD sothat the demand does not exceed the available oxygen. A mixed group of organisms should be present in the sample, if not, the sample has to be seeded artificially. Temperature is controlled at 20□C. The test is carried out for 5 days, as 70 to 80% of the BOD is satisfied during this period.

### *(ii) Interferences and Inadequacies*

The pH of the sample should be adjusted by adding alkali or acid, in the range 6.5 to 8.5.

Some samples may be sterile, and will need seeding. The purpose of seeding is to introduce into the sample a biological population capable of oxidizing the organic matter. Where such microorganisms are already present as in surface waters, domestic sewage or unchlorinated effluents etc., seeding is not required.

For seeding, wherever necessary use settled as to each litre of dilution water.

Other interference in BOD test are the same as in the DO test. In addition, lack of nutrients in dilution water, lack of acclimated seed organisms and presence of heavy metals or other toxic materials such as residual chlorine are other sources of interference in this test.

### *(iii) Apparatus*

*Incubation bottles* : 300 ml capacity with groundglass stoppers. New bottles should be cleaned with either 5 N hydrochloric or sulfuric acid followed by rinsing with distilled water. In normal use bottles once used for winkler procedures does not required any treatment apart from thorough rinsing with tap water and distilled water. As a precaution against air into the dilution bottle during incubation, use special 800 bottles and add water to the flared mouth of the bottle.

### *(iv) Reagents*

*Dilution water* : Distilled water alone is unsatisfactory as a diluent and it is recommended that a synthetic dilution water be employed. This can be prepared by adding reagents to good quality of distilled water. Water from copper stills should not be used since copper inhibits biochemical oxidation. Glass distilled or deionized water produced in some commercial units may be used.

*Stocks solutions of the following pure chemicals are required. Any solutions showing signs of precipitates or growth should be discarded.*

*Phosphate buffer stock solution* : Dissolve 42.5 g potassium dihydrogen phosphate ($KH_2PO_4$) in 700 ml water and add 8.6 g sodium hydroxide. This should give a solution of pH 7.2 which should be checked. Add 2 g ammonium sulphate and dilute to 1 litre.

*Magnesium sulfate solution* : Dissolve 25 g magnesium sulfate $MgSO_4.7H_2O$ in 1 litre water.

*Calcium chloride solution* : Dissolve 27.5 g anhydrous calcium chloride $CaCl_2$ in 1 litre distilled water.

*Ferric chloride solution* : Dissolve 0.125 g ferric chloride $FeCl_3.6H_2O$ in 1 litre water.

### *(v) Seeding*

The purpose of seeding is to introduce into the sample a biological population capable of oxidizing the organic matter in the waste water. Samples in which such microorganisms are already present, as in domestic sewage and surface waters seeding is not required.

When the sample contains very few microorganisms the standard seed material used is sewage either that has been stored at 20□C for 24 to 36 hrs. The seed added should produce a seed correction of atleast 0.6 mg/l. Usually 2 ml to 5 ml of sewage is used per litre of dilution water.

### *(vi) Procedure*

*Preparation of dilution water*

Aerate the required volume of distilled water in a container by bubbling compressed air for 12 days to attain DO saturation. After aeration keep it at 20□C for atleast on day.

At the time of use, add 1 ml each of phosphate buffer, magnesium sulphate, calcium chloride and ferric chloride for each litre of dilution water. The trend of adding these nutrients should be followed as given above. If required add requisite quantity of seed also. Mix well.

*Dilution of sample*

Neutrialize the sample to pH around 7.0 using alkali or acid (NaOH or $H_2SO_4$).

The sample should be free from residual chlorine. If it contains residual chlorine, remove it by using $Na_2SO_3$ solution as given below :

Take 50 ml of the sample and acidify with 10 ml of 1 + 1 acetic acid. Add about 1 g of KI. Titrate with Sodium sulfite using starch as indicator. Calculate the volume of $Na_2SO_3$ required per ml of the sample and add accordingly to the sample to be tested for BOD.

Atleast two dilutions of the sample are made so that the depletions are in the range 40 to 70%. It is recommended to have lower dilution to give 40% and higher one to give 70% oxygen depletion. The dilutions may be made as follows :

Requisite quantity of the sample is taken in a one litre capacity volumetric flask. Dilute this by siphoning out the dilution water to the mark. Mix well. Rinse BOD bottles with the sample and then fill the bottles with the diluted sample. As for as possible avoid entrapping of air bubbles in the bottle. Tap the neck of the bottles to ensure the removal of the air bubbles from the bottle. Stopper the bottle immediately.

Keep one bottle for determination of initial (zero day) dissolved oxygen and incubate two bottles at 20□C for 5 days. The bottles incubated should be water sealed. The water to seal the bottle should be added daily, as some water is lost due to evaporation.

Prepare four blanks by siphoning out only dilution water into the bottle. These bottles should be first rinsed and then filled with the dilution water. Two of these blanks should be used to determine initial dissolved oxygen and rest of the two are incubated for 5 days at 20□C. Usually depletion due to blank in 5 days should not be more than 0.2 mg/l.

*Dissolved oxygen, initial and after five days is determined as described in the chapter for the determination of the dissolved oxygen.*

*Seed correction* : If the dilution water is seeded, do not use the seeded blank for seed correction because the 5day seeded dilution water blank is subject to aeratic oxidation due to the very high dilution of seed, which is not the oxygen depletion of the seed by setting up a separate series of dilutions and selecting those resulting in 40 to 70% oxygen depletions in 5 days. Use one of these or average of these or average of these depletions to calculate the correction due to the small amount of seed in the dilution water.

### *(vii) Calculations*

(1) When BOD has been determined in an undiluted sample:

BOD mg/l = DO before incubation mg/l —
DO after incubation mg/l.

(2) When BOD has been determined in a diluted sample :

Definitions —

D1 = DO of diluted sample on 0th day

D2 = DO of diluted sample on 5th day

B1 = DO of dilution blank on 0th day

B2 = DO of dilution blank on 5th day

C1 = DO of seed control on 0th day

C2 = DO of seed control on 5th day

(3) F = Ratio of seed in sample to seed in control

$$F = \frac{\%\text{ seed in D1}}{\%\text{ seed in C1}}$$

Seed correction = (C1 – C2) F

(4) BOD without applying seed correction

$$\text{BOD mg/l} = \frac{(\text{D1} - \text{D2}) - (\text{B1} - \text{B2}) \times 100}{\%\text{ sample}}$$

(5) BOD applying seed correction

$$\text{BOD mg/l} = \frac{(\text{D1} - \text{D2}) - (\text{C1} - \text{C2}) \times 100}{\%\text{ sample}}$$

*Volume correction*

This correction is applied in case of low BOD samples were longer volume of sample is incubated. In such cases the fraction of dilution water used to dilute the sample is much less than 100% and hence depletion due to blank, for the added fraction only should be taken into account and not the 100% in normal cases.

The following modified equation is to be used for applying this correction :

*(1) BOD without applying seed correction*

$$\text{BOD mg/l} = \frac{(\text{D1} - \text{D2}) - (\text{B1} - \text{B2}) \times \text{A} \times 100}{\%\text{ sample}}$$

*(2) BOD applying seed correction*

$$\text{BOD mg/l} = \frac{(\text{D1} - \text{D2}) - (\text{C1} - \text{C2}) \times \text{F} \times \text{A} \times 100}{\%\text{ sample}}$$

Where A is the fraction of dilution water added and is given as

$$\text{A} = \frac{100 - \%\text{ Sample added}}{100}$$

For low percentage of samples A is unit.

***(viii) Expression of results***

BOD is expressed as mg/l, 5 days 20□C.

*Precision and Accuracy* : There is no acceptable procedure for determining the accuracy of the BOD test.

### *(ix) Precautions*

(1) The trend of adding nutrients should be phosphate buffer, magnesium sulphate, calcium chloride and ferric chloride.

(2) After aerating dilution water stablize it at 20□C for atleast one day.

(3) The pH of the sample should be in range 6.5 to 8.5.

(4) While siphoning the dilution water and filling BOD bottles avoid entrapping of bubbles.

(5) Wherever necessary apply seed and volume corrections.

(6) Add water on the neck of the bottles during incubation period to ensure continuous proper water sealing.

(7) Requisite treatment should be given to the samples containing interfering radicals like iron, residual chlorine, etc.

## 10. CHEMICAL OXYGEN DEMAND (COD)

### *(i) General*

The chemical oxygen demand determines the amount of oxygen required for chemical oxidation of organic matter using a strong chemical oxidant, such as potassium dichromate under reflux conditions.

The test is widely used to determine :

(1) the degree of pollution in water bodies and their self purification capacity.

(2) efficiency of treatment plants

(3) pollutional loads

(4) provides rough idea of BOD, which can be used for BOD estimation.

The limitation of the test lies in its inability to differentiate between the biologically oxidizable and biologically inert material.

COD determination has an advantage over BOD test in that the results can be obtained in less than 5 hours where as BOD determination requires 5 day. Further the test is relatively easy and precised. Also there are not many interferences as in the case of BOD.

### *(ii) Principle*

Most of the organic matter are destroyed when boiled with a mixture of potassium dichromate and sulfuric acid producing carbon dioxide and water. A sample is refluxed with a known amount of potassium dichromate in sulfuric acid medium and the excess of dichromate is titrated against ferrous ammonium sulfate. The amount of dichromate consumed is proportional to the oxygen required to oxidize the organic matter.

### *(iii) Interference*

Straightchain aliphatic compounds, aromatic hydrocarbons, fatty acids, chlorides, nitrites and iron are the main interfering radicals.

The interference caused by chlorides can be eliminated by the addition of mercuric sulfate to the sample prior to addition of other reagents. About 0.48 of $HgSO_4$ is adequate to complex 40 mg $Cl^-$ ions in the form of poorly ionized $HgCl_2$.

Nitrate exerts a COD of 1.1 mg/mg N. 120 mg of sulphuric acid is added to $K_2Cr_2O_7$ solution to avoid interference caused by $NO_2$. When 20 ml sample and 10 ml dichromate is taken, this can take care of concentrations up to 6 mg/l $NO^-_2N$.

Slider sulfate is added to conc. $H_2SO_4$ (22g/4kg acid) as a catalyst. This accelerates the oxidation of straight chain aliphatic and aromatic compounds.

For complete and better oxidation of organic matter it is necessary to see that the final concentration of $H_2SO_4$ cocn. is 50%.

### *(iv) Apparatus*

Reflux Apparatus consisting of a flat bottom, 250 to 500 ml capacity flask with ground glass joint and a condenser with 24/40 joint.

*Hot plate* : It is convenient to have a bigger hot plate on which a series of refluxing sets is connected to have a single water circulation to all. This also eliminates the botheration of having separate blank for each sample.

### *(v) Reagents*

*Standard Potassium dichromate, 0.25 N* : Dissolve 12.259 g potassium dichromate previously dried at 103°C for 24 hrs in distilled water. Add about 120 mg sulpemic acid to this. Dilute to 1 litre.

*Sulpemic acid* : Add 22 g silver sulfate to 9 lbs (4kg) conc. $H_2SO_4$ bottle. Keep overnight for dissolution. Shake well after dissolution.

*Standard Ferrous Ammonium Sulfate, 0.1 N*

Dissolve 39 g $Fe(NH_4)_2(SO_4)_2.6H_2O$ is distilled water. Add 20 ml conc. $H_2SO_4$, cool, and dilute to 1 litre. Standarize this solution daily against the standard $K_2Cr_2O_7$.

## Standarization

Dilute 10.00 ml standard $K_2Cr_2O_7$ solution to about 100 ml. Add 30 ml conc. $H_2SO_4$ and cool. Add 3-4 drops of ferroin indicator and titrate with ferrous ammonium sulfate till the colour changes to wine red.

$$\text{Normality of Ferrous Ammonium Sulfate} = \frac{10 \times 0.25}{\text{ml Fe } (NH_4)_2\ (SO_4)_2}$$

*Ferroinindicator*

Dissolve 1.485 g 1,1-Dpheninthroline monohydrate, and 695 mg $FeSO_4.7H_2O$ in water and dilute to 100 ml.

This indicator solution may be purchased already prepared.

Mercuric Sulfate, $HgSO_4$ Crystals

### *(vi) Procedure*

Place 0.4 g $HgSO_4$ in a reflux flask. Add 20 ml or an aliquot of sample diluted to 20 ml with distilled water. Mix well, sothat chlorides are converted into poorly ionized mercuric chloride. Add 10 ml standard $K_2Cr_2O_7$ followed by slowly 30 ml sulfuric acid which already contains silver sulfate. This follow addition alongwith swirling prevents loss of volatile materials such as fatty acids, in the sample. Mix well if the colour turns green either take fresh sample with smaller aliquot or add more dichromate and acid. Final concentration of conc. $H_2SO_4$ should be always 50%.

Connect the flask to condensers and reflux for 2 hours. Cool and wash down the condensers with small quantity of distilled water. Remove the flask and add about 80 ml distilled water. Cool and titrate against standard ferrous ammonium sulfate using ferroin as indicator. Colour changes sharply from green-blue to wine red.

Reflux a reagent blank under identical conditions preferably simultaneously with the sample.

Alcornate procedure for low-COD samples (below 50 mg/l)

The method is same as for the high-COD samples except in this case the concentrations of $K_2Cr_2O_7$ and ferrous ammonium sulfate is reduced i.e. $K_2Cr_2O_7$, 0.05N and ferrous ammonium sulfate, 0.025N. Exercise extreme care with this procedure because even a trace of organic matter in the glassware or the atmosphere may cause a gross error. COD values upto 160 mg/l can be estimated by this procedure.

***(vii) Calculation***

$$\text{mg/l COD} = \frac{(a - b)\, N \times 8000}{\text{ml sample}}$$

Where,

a is the ml $Fe(NH_4)_2\ (SO_4)_2$ required for blank

b is the ml $Fe(NH_4)_2\ (SO_4)_2$ required for sample

N is the normality of $Fe(NH_4)_2\ (SO_4)_2$

***(viii) Expression of Results***

Results are expressed as mg/l of $O_2$

Report to the nearest whole number.

***(ix) Precision and Accuracy***

Precision and accuracy both depends upon the COD value.

For the sample having COD higher than 400 mg/l precision upto 2% is expected from a good analyst. As the COD value goes on decreasing precision also become proper and poorer i.e. % goes on increasing.

Precision for the low COD samples may be improved by using alternate method where diluted reagents are used.

The accuracy of this method is till not very clear. For most organic compounds the oxidation is 95 to 100% of the theoretical value.

***(x) Precautions***

(1) The strength of the sulfuric acid in final solution should be 18 N.

(2) The trend of making mixture should be $HgSO_4$, sample (swirl), $K_2Cr_2O_7$ and conc. $H_2SO_4$ (slowly with swirling).

(3) As for as possible reflux blank and samples simultaneously.

(4) After refluxing, cool, use required amount of distilled water for washing the condenser. Cool and then titrate.

## 11. NITROGEN

Nitrogen content in sewage water can be determined by nesselarization method as suggested by Mc. Donald (1978).

### *Procedure*

One ml of sewage water is taken for digestion in corning test tube which is digested with few drops of conc. $H_2SO_4$ and a pinch of digestion mixture (The digestion mixture consist of potassium sulfate, sodium triosulfate, salicylic acid and selenium), after cooling few drops of 30% hydrogen peroxide is added and heating is continued for some time. Finally the colourless fluid is made up with distilled water to volume 5 ml, and to this 0.2 ml of sodium silicate (10%) and sodium hydroxide (10%) are added and shake thoroughly, the final volume is made to 10 ml, with distilled. Then, 2 ml of diluted (2 : 1) nesseler's reagent is mixed and shaked. The absorbance of yellow colour after 10 minutes is read at 425 nm. by spectrophotometer. The amount of nitrogen in samples is calculated with the help of standard curve, prepared from ammonium sulfate and results are expressed as μg nitrogen/ml of sewage water.

## 12. SULFATE

***(i) General*** : Sulfate is widely distributed in nature and may be present in natural waters in concentrations ranging from a few to several thousand milligrams per litre. Mine drainage wastes may contribute high sulfate by virtue of pyrite oxidation. Sodium and magnesium sulfate exert cathatic action and hence its concentration above 250 mg/l in potable water is objectionable. Sulfates cause a problem of scaling in industrial water supplies, and problem of odour and corrosion in waste water treatment due to its reduction to $H_2S$.

Though the gravimetric method is most accurate for sulfate concentration above 10 mg/l, natural waters are likely to have low concentrations and hence turbidimetric method which is more rapid and also is more or less equally accurate as the former method, is suggested. Turbidimetric method may be applied to the concentrations up to 60 mg/l. When the concentration exceeds 60 mg/l it may be diluted accordingly.

***(ii) Principle*** : Sulfate ions are precipitated in a hydrochloric acid medium with barium chloride so as to form a uniform suspension of $BeSO_4$ crystals. The absorbance of the suspension is measured by a photometer and the sulfate concentration is determined by comparison of the reading with a standard curve.

***(iii) Interference*** : Colour and suspended matter interfere in determination. Some suspended matter may be removed by filtration. If both are small in comparison to sulfate on concentration, this interference is corrected by running blanks in which $BaCl_2$ is eliminated. Silica in excess of 500 mg/l will interfere.

There are no ions other than sulfate in normal waters that will form insoluble compounds with barium under strongly acid conditions.

*Minimum detectable concentration* : Approximately 1 mg/l sulfate.

***(iv) Apparatus***

(a) *Magnetic Stirrer* : It is better to use a magnetic stirror to obtain constant stirring rate and time (usually one minute is sufficient).

(b) *Photometer* : One of the following is required.

(1) Nelholometer

(2) Spectrophotometer (at 420 nm)

(3) Filter photometer (at 420 nm, violet blue filter)

(c) Stopwatch.

***(v) Reagents***

*Conditioning reagent* : Mix 50 ml glycerol with a solution containing 30 ml Conc. HCl, 300 ml distilled water, 100 ml 95% ethyl alcohol and 75 g NaCl.

*Barium Chloride* $BaCl_2$ **:** Crystals, 20-30 mash.

*Standard Sulfate Solution* : (a) Dissolve 1.4798 anhydrous $Na_2SO_4$ in distilled water and dilute to 1000 ml.

Dilute 100 ml of solution "a" to 1 litre. This solution is 1 ml = 100/μg $SO_4$.

***(vi) Procedure*** **:** Take 100 ml conical flask. Add exactly 5 ml conditioning reagent and mix it on the magnetic stirror. Add a spoonful of $BaCl_2$ crystals while stirring. Continue stirring for exactly 1 min. at a constant speed.

Immediately after the stirring is over, four some samples into the absorption cell of the photometer and measure the turbidity after every 30 sec. for 4 minutes at 420 nm (violet blue filter). Maximum turbidity usually occurs within 2 min. and the reading remains constant thereafter for 3 to 10 min.

Prepare standard curve by carrying standard sulfate solution through entire procedure. Space standards at 6 mg/l increment in the range 0 to 40 mg/l. Run reagent blank under identical conditions, it is convenient to adjust the photometer (either 100% transmission or zero absorbance) with this blank before taking the sample readings.

To account for the interference due to the colour and turbidity of the sample, run sample blank in which only $BaCl_2$ is withstand. For a particular sample either this may be used to adjust the instrument or the absorbance due to this is deducted from the sample absorbance. Use same amount of sample for the sample blank and sample determination.

***(vii) Calculation***

$$\text{mg/l } SO_4 = \frac{A \times 1000}{\text{ml sample}}$$

Where A is the concentration of sulfate ion in mg, obtained from the calibration curve for the corresponding sample absorbance.

***(viii) Expression of results***

Results are expressed as mg/l $SO_4$.

***(ix) Precautions***

(1) Time and rate of stirring should be kept constant throughout the determination i.e. for standards, blank and samples.

(2) Final volume should be kept constant.

(3) Make all turbidity measurements after a constant selected time, when the turbidity is maximum.

(4) Wherever required use sample blanks to account for the interference due to turbidity and colour of the sample.

Suspended matter may be removed by filtration.

(5) Method holds good up to the concentration 60 mg/l. For higher concentration, dilute proportionately.

## 13. MAGNESIUM

***(i) General*** **:** Magnesium is a common constituent of natural water. It is one of the important contributors to the hardness of water. Magnesium salts broke down on heating to form scale in boilers. Concentrations greater than 125 mg/l can also exert a cathartic and diuretic action. The magnesium concentration in water may vary from zero to several hundred milligrams per litre, depending on the source of the water.

Chemical softening of ion exchange reduces the magnesium and associated hardness to acceptable levels.

***(ii) Magnesium by calculation***

For routine estimation of magnesium concentration in water, the magnesium is calculated from the EDTA calcium and total hardness titration methods.

Magnesium = Total hardness – Calcium hardness

(all expressed as mg/l $CaCO_3$)

## 14. CALCIUM

***(i) General*** **:** The presence of calcium in surface water results from passage through or over deposit of lime stone, dolamite, gypsum and gypsiforous shale. The concentration of calcium may range from zero to several hundred miligram per litre depending on the source. Small concentrations of calcium carbonate prevents corrosion of metallic pipes by laying down a protective coating. On the other hand, appreciable quantity of calcium salts break down on heating, to form harmful scale in boilars, pipes and cooking utensils. Calcium contributes to the total hardness of water.

The atomic absorption spectrophotometer method is the most accurate method available for the calcium determination. But the EDTA and permagnate titrimetric methods also give good results for control and routine applications. Due to simplicity and rapidity of the EDTA method it has become first choice of the analyst. Hence, only EDTA titrimtric has been described.

***(ii) Principle*** : When EDTA is added to water containing both calcium and magnesium it combines first with the calcium. By titrating with EDTA calcium can be determined directly, when the pH is sufficiently high sothat the magnesium is precipitated as the hydroxide and an indicator is used which combines only with the calcium. Indicator shows a colour change when all of the calcium is complexed by EDTA, at a pH of 12 to 13.

***(iii) Interference*** : Under conditions of test, the following concentrations of the ions causes no interference with the calcium hardness determination :

| | |
|---|---|
| Copper 2 mg/l | Zinc 5 mg/l |
| Ferrous iron 20 mg/l | Lead 5 mg/l |
| Ferric iron 20 mg/l | Aluminium 5 mg/l |
| Manganese 10 mg/l | Tin 5 mg/l |

Orthophosphate will precipitate calcium at the pH of test. Stroncium and barium also interfere. Alkalinity in excess of 30 mg/l may cause an indistinct and point with hard waters.

### *(iv) Reagents*

*Standard EDTA titrant, 0.01 M* : Dissolve 3.723 g analytical reagent grade ethylenediminetetraacetic acid disodium salt. $Na_2H_2C_{10}H_{12}O_8N_2.2H_2O$in distilled water, and dilute to 1 litre. Standarize against standard calcium solution. This solution is approximately 1.00 mg $CaCO_3$/ ml.

*Standard Calcium Solution, 0.01 M* : Weigh 1 g anhydrous calcium carbonate, $CaCO_3$ power into a 500 ml erlenmeyer flask. Place a funnel on the flask and add slowly HCl until all the $CaCO_3$ has dissolved. Add 200 ml distilled water and boil for few minutes to expel out $CO_2$, cool, add few drops of methyl red indicator, and adjust to the intermediate orange colour by adding 3N $NH_4OH$ or 1 + 1 HCl, as required. Make up the volume up to 1 litre in a volumetric flask. This standard solution is equivalent to 1 mg $CaCO_3$/1ml.

### *(v) Indicator*

Mureoxide (ammonium purpurate) was the first indicator available for the detection of the calcium end point, but due to difficulty in recognizing the end point it is rarely used these days. Eriochrome Blue Black R or Solochrome Dark Blue (also called Calcon) gives an improve and point, changing colour from red to pure blue.

*Eriochrome Blue Black R* : Prepare a stable form of the indicator by grinding together in a mortar 200 mg powdered dye and 100 g solid sodium chloride to a 40 to 50 mash. Store in a tightly stoppared bottle. Use about 0.2 to 0.4 g of the mixture for each titration. During the course of titration the colour changes from red through purple to bluish purple tint. The pure blue with out any trace of reddish or purple tint. The pH of the some sample (not all) must be raised to 14 (rather than 12 to 13) by using 8N NaOH in order to get a sharper colour change.

*Sodium hydroxide, NaOH, N* : Dissolve 40 g sodium hydroxide in distilled water, and make up the volume to 1 litre. Store in polythene bottle.

### (vi) Procedure

Take 100 ml or smaller portion diluted to 100 ml so that calcium content is about 10 to 20 mg. Add 5 ml NaOH solution to produce a pH of 12 to 13. Stire and add about 0.2, 0.4 g of indicator mixture. Because of the high pH required, titrate immediately after adding alkali and indicator. Add EDTA titrant slowly with continuous striking to the proper and point.

Carry out a reagent blank using 100 ml distilled water under identical conditions.

Standarize EDTA titrant with 25 ml of calcium standard. Follow the same method as described for the sample.

### (vii) Calculation

$$\text{Calcium as mg/l } CaCO_3 = \frac{(V_1 - V_2) \times 1000}{\text{ml sample}} \times \frac{25}{V_3}$$

Where,

$V_1$ is the volume of EDTA required for sample.

$V_2$ is the volume of EDTA required for reagent blank.

$V_3$ is the volume of EDTA required for 25 ml of standard calcium.

*Expression of results* : Results are expressed as mg/l of $CaCO_3$. Round off the results to the nearest whole number.

### (viii) Precision and Accuracy

The precision and accuracy of about ± 2 to 4 % is expected from a good analyst, when the concentration is more than 25 mg/l. Precision and accuracy also depends on the quantity of calcium present.

### *(ix) Precautions*

(1) Titrate immediately after adding the indicator because it is unstable under alkaline conditions.

(2) For reconising end point use a colour comparison blank prepared under identical conditions.

(3) The final volume for sample and blank should be kept constant.

(4) Use deionized distilled water for solution preparations, and any dilution necessary.

(5) Add same quantity of indicator to sample and blank, so that the colour comparison is better.

## 15. SODIUM AND POTASSIUM

Sodium and Potassium contents of sewage water are determined by flame photometer as suggested by APHA (1975).

### *Procedure*

10 ml of well mixed sewage water is taken and it is digested with conc. nitric acid till white residue on drying is left. After that, samples are dissolved in conc. HCl and warm distilled water and after filtration the final volume is made upto 10 ml. This solution is atomized in flame photometer and the concentration of sodium and potassium are determined with the help of calibration curves and the results are expressed in ppm.

# PART – III
# METHODOLOGY FOR SOIL ANALYSIS

**PART – III**

# METHODOLOGY FOR SOIL ANALYSIS

## 1. SOIL pH

The glass electrode pH meter is standardized with standard pH 4 to 9 solutions. After standarization the electrode is washed with distilled water and then immersed into the soil suspension of 1 : 5 (Soil : Water) and keep for four hours. The pH of the clear suspension is determined.

## 2. SOIL CONDUCTIVITY

Soil conductivity is measured with the help of conductivity meter by preparing 1 : 5 (Soil : Water) soil suspension in deionized water after keeping for four hours.

## 3. SOIL MOISTURE

Soil samples are collected from experimental site at the depth of 0.5 and 50-60 cm (may be vary), packed in polythene bags and are brought to laboratory. Fresh and oven dried (at 108°C for 24 h) weight of each samples are determined. The percentage moisture content of the soil is determined as dry weight basis as follows :–

$$\text{Moisture \%} = \frac{\text{Fresh weight} - \text{Dry weight}}{\text{Dry weight}} \times 100$$

## 4. SOIL DENSITY

Soil samples are oven dried at 105°C until a constant weight is sustained. The dried sample is transferred to measuring cylinder and the volume is noted. The weight of this volume of the soil is calculated to determine the density of the soil sample :

$$\text{Soil Density} = \frac{\text{Weight of Soil}}{\text{Volume of Soil}}$$

## 5. PORE SPACE OF SOIL (Narayana and Shah, 1966)

The Pore Space of the Soil is determined by the following formula :

$$\text{Pore Space} = \frac{\text{Apparent specific gravity}}{\text{True specific gravity}} \times 100$$

Where apparent specific gravity is obtained by dividing the air dried weight of a soil by the volume. While true specific gravity of the soil is 2.65.

## 6. WATER HOLDING CAPACITY OF SOIL

2 mm sieved fresh soil is taken in a 50 ml glass specimen tube and the mouths are closed with filter papers by a rubber band. The tube containing soil is immersed in a large beaker having distilled water 2 cm of depth. After 24 hours, the filter papers are removed carefully and the tubes are weighed, then it is kept in oven at 105□C for 24 hours, and reweigh.

$$\text{Water Holding Capacity (\%)} = \frac{W2 - W3}{W3} \times 100$$

Where,

W2 = Weight of soil after 24 hours, water saturation.

W3 = Weight of oven dry soil.

## 7. CATION EXCHANGE CAPACITY OF SOIL (Bower et al., 1952)

Cations retained on negatively charged soil particales by weak forces are termed as exchangeable and the amount of these cations is termed as cation exchange capacity and the amount of individual exchangeable cations are usually expressed in milliequivalents (meq) per 100 g of soil. The exchange capacity of soils ranges from 5 to 50 meq/100 g, and is greater in organic soils. The absorbed cations interchanged with cations in the soil solution according to the law of mass action.

$CaX_2 + 2\,Na^+ = 2\,NaX + Ca^{2+}$

Where, X denotes the exchange site.

### *(i) Reagents*

(A) *Sodium acetate solution (1N)* : Dissolve 136 g of sodium acetate in water and dilute to a volume of one litre. The pH of solution is made approximately to 8.2.

(B) Ethanol 95%

(C) *Ammonium acetate solution (1N)* : Dissolve 57 ml of conc. acetic acid to 800 ml of distilled water and to this 65 ml of conc. ammonium hydroxide is added. The solution is diluted to a volume of one litre and adjusts to pH 7.0 by the addition of ammonium hydroxide or acetic acid.

### *(ii) Procedure*

To the 4 g of air dried soil, 33 ml of reagent (A) is added and place in a centrifuge tube and shake for 5 min. The supernatant liquid is discarded and treat the sample in this manner with 33 ml of Reagent (A), a total of 4 times, discarding the supernatant each time. Again to this, a volume of 33 ml of Reagent (B) is added and centrifuge until the supernatant liquid is clear. Discard the supernatant liquid again and wash the soil with 33 ml of Reagent (B), a toatl of three times. The electrical conductivity of the supernatant liquid from the third washing is less than 40 micromhos/cm. Replace the absorbed sodium from the sample by extraction with three 33 ml portions of Reagent (C) and determine the sodium concentration of the combined extracts after making dilution to 100 ml. The values of the sodium from the samples are predicted by the help of a standard graph of sodium and the CEC is calculated by using the following formula

$$\text{CEC (meq/100 g)} = \frac{(\text{Na Conc. of extract in meq/l} \times 10)}{\text{Weight of soil sample in gram}}$$

## 8. FIELD CAPACITY OF SOIL (Barger, 1922)

Pots containing air dried soil samples are allowed to saturate in water bath for 24 hours. The pots are kept out of water for 14 hrs. So that the excess of gravitationalwatermoves down. The percentage water content is calculated as given below :

$$\text{\% of Water Content} = \frac{\text{Fresh weight– Dry weight}}{\text{Dry weight}} \times 100$$

## 9. ORGANIC CARBON IN SOIL

The nature and the amount of organic carbon greatly influence the site quality, soil structure, water holding capacity, availability and abundance of nutrients and biological populations and their activities.

Determination of organic matter as organic carbon is mostly widely used and some factors has been worked out by multiplying with which the value of organic carbon is converted into organic matter. The general factor most frequently used is 1.724. But since the nature of organic matter in different soils is very complex, the use of general factor should be avoided and valve obtained for organic carbon should be reported as such either by Method I or by Method II.

## Method – I

### *(i) Reagents*

All reagents should be of analytical reagent (AR) grade.

*$K_2Cr_2O_7$ solution* : Dissolve 49.0 g $K_2Cr_2O_7$ in distilled water and make to 1000 ml; it is not necessary to dry the $K_2Cr_2O_7$.

Conc. $H_2SO_4$

20 % v/v $H_2SO_4$

Sucrose standards. Dissolve 11.8755 g sucrose in distilled water and make to 100 ml (=50 mg C/ml). Dilute this solution with distilled water to produce standards in the range 2.5-12.5 mg C/ml.

Standards should be freshly made for each batch of analysis.

### *(ii) Apparatus*

Technicon Auto-Analyzer.

Log Chart Recorder Paper.

Hot plate

2 ml pipettes

10 ml pipettes with a safety bulb

25 ml and 100 ml measuring cylinders

1000 and 100 ml volumetric flasks

200 ml conical flasks

### *(iii) Procedure*

Dry soil samples at 105°C overnight and then grind finely in the disc mill. Between 0.2 and 2.0 g soil, depending on C-content, is weighed into a 250 ml conical flask and 10 ml potassium dichromate solution added from a pipette fitted with a safetybulb, followed by 20 ml conc. $H_2SO_4$ added quickly from a measuring cylinder. Mix thoroughly and stand on a hot plate for 5 minutes. Heating for longer than 5 minutes will result in decomposition of the dichromate. Cool and add 70 ml distilled water. Stand overnight to allow particles to sediment then decant the clear solution for analysis – Do not filter.

A standard calibration curve is prepared by treating 2 ml samples of standard solutions as described above. A blank of 10 ml potassium dichromate solution and 20 ml conc. $H_2SO_4$ is also prepared.

*N.B.* : Run 20% $H_2SO_4$ through the Auto Analyzer for atleast 1 h before running the samples (sample cup volume 8-10 ml). Standard solutions should be run first followed by 2 washes of 20% $H_2SO_4$ and then the samples. A wash of 20% $H_2SO_4$ should be run after every tenth sample. The Auto Analyzer is run at 40 samples/h with a sample : wash ratio of 2:1.

### *(iv) Calculation of results*

Draw a calibration curve of mg C against optical density. The standards obey Beer's law, but do not go through the zero as the dichromate gives a small reading. Calculate the slope of the line and use this to calculate mg C in each sample as follows :

$$\text{mg C} = (\text{peak height} - \text{Blank}) \times \text{slope}$$

## Method – II

Walkey and Black's titration method (Piper, 1944) can also be followed to determine the organic carbon in soil sample. In this method results are expressed in percentage.

### *(i) Preparation of reagents*

1. *N Potassium dichromate* : Dissolve 49.04 g of pure potassium dichromate in distilled water and dilute to one litre.
2. *Sulphuric acid* : Dissolve 1.25 g of silver sulphate per 100 ml of conc. sulphuric acid.
3. 85 % Phosphoric acid.
4. *Ferrous sulphate solution* : Dissolve 2.190 g of pure $FeSO_4.7H_2O$ in water and 15 ml of conc. $H_2SO_4$ and dilute

to one litre. Titrate this against N potassium dichromate and standarise.

5. *Diphenylamine indicator* : Dissolve 0.5 g of pure diphenylamine in a mixture of 100 ml of conc. $H_2SO_4$ and 20 ml of distilled water.

### *(ii) Method*

Transfer 1-5 g of soil which is dried and powdered, screen through 0.5 mm sieve plate into a 500 ml flask. To it add 10 ml of N Potassium dichromate and 20 ml of distilled water, 10 ml of 85 % phosphoric acid and 1 ml of diphenylamine indicator and it is run in N ferrous sulphate slowly with a constant stirring through a burette until the solution is purple or blue and continuously add ferrous sulphate solution until the colour become greenish. Add 0.5 ml of N Potassium dichromate and complete the titration by adding more ferrous sulphate, until the last trace of coulour disappear.

Since one ml of N Potassium dichromate corresponds to 3 mg of carbon, the percentage of organic carbon can be calculated by the following expression.

$$\% \text{ Organic Carbon} = \frac{V_1 - V_2 \times 0.003 \times 100}{W}$$

Where,

$V_1$ = ml of N $K_2Cr_2O_7$ (10.5 ml)

$V_2$ = ml of N Ferrous sulphate

W = Weight of the soil taken.

## 10. SODIUM AND POTASSIUM IN SOIL (Jackson, 1962)

Sodium and Potassium concentration of the soil can determined by flame photometer. 100 mg of air dried soil is digested with 1 ml of 2:1 nitric perchloric acid mixture and it is diluted to 10 ml with distilled water.

The samples are atomized in flame photometer and with the help of calibration curves the concentration of sodium and potassium are determined. The results are expressed in percentage.

## 11. TOTAL NITROGEN IN SOIL

### Method -I

#### *(i) Reagents*

Kjeldahl copper catalyst tablets

Kjeldahl selenium catalyst tablets

Conc. $H_2SO_4$, Nitrogen free

40% w/v NaOH solution

0.005 M Standard $H_2SO_4$

4% w/v $H_3BO_3$ solution

#### *(ii) Apparatus*

Kjeldahl digestion stand

30 ml Kjeldahl flasks

Semi-micro Kjeldahl distillation apparatus

Automatic titrator and pH meter

80 mesh sieve

3 ml safety pipettes

20 ml pipettes

50 ml volumetric flasks

#### *(iii) Procedure*

Air dry the soil and grind to pass an 80 mesh sieve. Weigh 1 g ground soil and place in a 30 ml Kjeldahl flask. Add approximately 1 ml water and when this has wetted the soil add 3 ml conc. $H_2SO_4$ with a safety pipette. Gently boil off the water, then add half a copper catalyst tablet and half a selenium catalyst tablet. Transfer to the Kjeldahl digestion stand and heat strongly until the liquid clears. if solid adheres to the walls of the flasks wash down with approximately 1 ml conc. $H_2SO_4$. Continue heating for a further 4 h. Cool, dilute with water, pour into a 50 ml volumetric flask together with the washings and make upto the mark. Do not filter.

Transfer 20 ml solution into the distillation apparatus, add approximately 10 ml of 40% NaOH and steam distil into 5 ml of 4% $H_3BO_3$. Titrate the distillate to pH 5.2 with 0.005 M $H_2SO_4$ using the automatic titrator.

### *(iv) Calculation*

$$\% \text{ N in dry soil} = \frac{T \times 0.014}{W}$$

Where,

T = Titre (ml)

W = Weight of dry soil.

### Method-II – (Mc Donald, 1978)

10 mg of oven dried soil is digested in a corning test tube with 12 drops of conc. $H_2SO_4$ and a pinch of digestion mixture which consist of following chemicals -

| | |
|---|---|
| Potassium sulphate | 2.0 g |
| Sodium thiosulphate | 1.5 g |
| Salicylic acid | 1.5 g |
| Selenium | 1.0 g |
| Total | 5.0 g |

After cooling the digestion mixture, few drops of hydrogen peroxide is mixed and heat for some time. Finally the colour less digested fluid is made to 5 ml and then add 0.2 g of sodium silicate and NaOH (10%). After thorough shaking the volume is made to 10 ml with distilled water. 2 ml of Nessler's reagent is mixed and the intensity of the yellow colour is read at 425 nm. The amount of nitrogen is calculated from standard curve prepared by ammonium sulphate.

## 12. AVAILABLE PHOSPHATE IN SOILS

### *(i) Reagents*

0.05% w/v aqueous polyacrylamide, M.W. > $5 \times 10^6$

0.5 M $NaHCO_3$ solution : Dissolve 210 g $NaHCO_3$ in deionized water, add 25 ml 0.05% w/v aqueous polycrylamide solution and make upto 5 l. Adjust to pH 8.5 with 50 % NaOH solution.

0.6% w/v $(NH_4)_6Mo_7O_{24}$ $4H_2O$ solution

Dissolve 30 g powedered $(NH_4)_6Mo_7O_{24}$ $4H_2O$ and 0.75 g antimony potassium tartrate in 3 l. deionized water. Slowly add 370 ml conc. $H_2SO_4$ and dilute to 5 l. Store in a dark bottle at 4□C. 0.15% w/v $(NH_4)_6Mo_7O_{24}$ $4H_2O$ solution.

Dilute 250 ml 0.6 % $(NH_4)_6Mo_7O_{24}$ $4H_2O$ solution to 1 l. Store in a dark bottle at 4□C.

1.5 % w/v ascorbic acid solution : This reagent must be made up on the day of use.

Approx. 1.5 M $H_2SO_4$

Add 256 ml 1.5 M $H_2SO_4$ to 1 l. 0.5 M $NaHCO_3$ solution. Stir vigorously on the magnetic stirrer for 15 min to remove all $CO_2$.

***(ii) Standard Solution***

Dissolve 0.879 g $KH_2PO_4$ (Dried at 105□C for at least 2 h) in 150 ml deionized water. Add 1 ml conc. $H_2SO_4$, make up to 200 ml and add 1 drop toluene. 5 ml of this solution is diluted to 250 ml with 0.5 M $NaHCO_3$ to give a 20 μg P/ml solution for preparation of standards. Standards in the range 1-7 μg P/ml are made by diluting the appropriate volumes of the 20 μg P/ml solution to 100 ml with 0.5 M $NaHCO_3$.

***(iii) Apparatus***

Technicon Autoanalyzer

pH meter

2, 5, 10 and 25 ml pipettes

100, 250, 1000 and 5000 ml volumetric flasks

50 ml automatic dispenser

500 ml measuring cylinder

250 ml bottles

End-over-end shaker

Filter funnels, 7 cm diam

Whatman No. 2 filter papers

100 ml conical flasks

***(iv) Procedure***

Weigh 5 g air dry soil, ground to pass a 2 mm mesh sieve, into 250 ml bottles and add 50 ml 0.5 M $NaHCO_3$ with the automatic dispenser. Shake for 45 min and then filter through Whatman No. 2 filter papers. Pipette 10 ml of the extract into a 100 ml conical flask, add 2ml 1.5 M $H_2SO_4$, swirl to release $CO_2$ and allow to stand overnight. Transfer to Auto Analyzer cups and analyse at 40 samples/h with a 2:1 sample to wash ratio.

### (v) Calculation

The calibration curve does not obey Beer's law.

μg P/g dry soil = μg/ml (from calibration curve) x 10

## 13. DETERMINATION OF RESPIRATION AND CARBONDIOXIDE EVOLUTION

### (i) Reagents

0.5 M $H_2SO_4$

0.5 M $H_2SO_4$ prepared from a concentrated volumetric solution. M NaOH prepared by dilution from a 50 % (w/v) solution. This reduces contamination by carbonates which are present on NaOH pellets and which participate from the concentrated solution.

Molecular sieve, type 5A 1/8 inch pellets.

Silica gel : Soda asbestos, 6-12 mesh.

$CO_2$-free water. (Distilled water boiled for 1h and stored in a bottle fitted with a soda-asbestos trap).

### (ii) Apparatus

Scrubbing system

Respirometer flask units

Automatic titrator and pH meter

50 ml automatic dispenser

50 ml pipettes

100 ml volumetric flasks

125 ml Drechsel bottles

### (iii) Procedure

Samples (100 g) of control or treated soil are placed in respirometer flasks and connected to the gas flow lines from the scrubbing system. Gas flow rate is controlled largely by the capillary tube but a fine adjustment to 1-2 bubbles/sec in the Drechsel wash bottle is obtained by moving the adjustable inlet pipe through the wash bottle head.

In general experience 4 replicates of each treatment and 4 blanks (Flasks with no soil) are sufficient.

NaOH (40 ml) is placed in the Drechsel bottles using an automatic dispenser. The bottles are replaced at 1 or 2 week intervals and the NaOH titrated to determine dissolved $CO_2$. The titration can be done manually but an automatic titrator is desirable as it allows larger numbers of samples to be handled.

The NaOH from the Drechsel bottle is made to 100 ml in volumetric flasks with $CO_2$-free water and 50 ml then pipetted into the titration vessel of the automatic titrator. The solution is adjusted to approximately pH 9.5 with 0.5 M $H_2SO_4$ and then to pH 8.3 with 0.05 M $H_2SO_4$. The volume of 0.05 M $H_2SO_4$ required to titrate this solution between pH 8.3 and 3.8 is noted.

***(iv) Calculation of results***

Carbon absorbed as $CO_2$ is calculated as :

$$\text{ppm C as } CO_2 = \frac{[T - B] \times 12 \times 2 \times 100}{100 - M}$$

Where,

T = titre of sample (ml)

B = titre of blank (ml)

M = moisture content of soil (%)

## 14. DETERMINATION OF OXYGEN UP-TAKE

***(i) Reagents***

M NaOH

0.2 % (w/v) glucose solution

***(ii) Apparatus***

*Gilson Respirometer* : Respiration flasks are standard Cuickfit 50 ml Erlenmeyer flasks with central wells (2.5 ml) fixed to the base.

2 ml graduated pipettes.

***(iii) Procedure***

Equilibrate the respirometer at the required temperature, normally 20▯C. Place 20 g treated or control soil in the respiration flasks taking care not to get any in the central well. A fluted filter paper wick is placed in the well of each flask so that it protrudes by 5 mm and 0.5 mm and 0.5 ml M NaOH put in each well. The flasks are then fitted to the respirometer with the appropriate reduction adaptors and allowed to equilibrate for at least 30 min.

After eqilibration the taps on the respirometer are opened and readings taken every 30 min for 4-5 h.

With soils having low oxygen uptake rates 2.0 ml of 0.2 % (w/v) glucose solution can be added to increase uptake. Care must be taken in interpreting results obtained by this method as the glucose treatment can seriously modify the soil microflora.

### *(iv) Calculation of Results*

Oxygen uptakes are calculated as :–

$$\mu l\ O/g\ soil/hour = \frac{100\ R \times \dfrac{273\ p}{760\ (t + 273)}}{W\ (100 - M)\ T}$$

Where,

R = uncorrected O uptake readings (ul)

p = barometric pressure (mmHg)

t = temperature of respirometer water bath (□C)

W = Wet weight of sample (g)

M = Moisture content of soil (%)

T = Time (h)

## 15. DETERMINATION OF SOIL ENZYME : DEHYDROGENASE

### *(i) Reagents*

3% (w/v) 2,3,5-triphenyl tetrazolium chloride (TTC) solution. Dissolve 0.3 g TTC and 0.6057 g tris-(hydroxymethyl)-methylamine in 10 ml deionized water. This solution should be made up immediately before use and kept in the dark.

### *(ii) Methanol*

*Standard solutions* : Dissolve 10 mg 2,3,5-triphenyl tetrazolium formazan (TPF) in 1 l. methanol. This solution (10 ug TPF/ml) is stable if stored in the dark at 4□C. Working standards are prepared by diluting this stock solution with methanol to give TPF concentrations in the range 2 to 8 ug TPF/ml.

### *(iii) Apparatus*

Spectrophotometer

Vortex mixer

End-over-end shaker

Centrifuge

25 ml volumetric flasks

15 ml screw-cap specimen tubes

0.5 ml graduated pipettes

### *(iv) Procedure*

Samples (1g) of treated or control soils are weighed into the screw-cap specimen tubes and 0.2 ml TTC solution and 0.5 ml deionized water added. The soil suspensions are then mixed for 15 seconds on the vortex mixture and incubated in the dark at 30□C for 18 h.

*N.B.* : It is important that subsequent stages in the procedure are done in subdued light to avoid fading of the colour of the TPF solutions.

Add 10 ml of methanol to each tube and shake on the end-over-end shaker for 30 min. Centrifuge the extracts at 2000 g for 20 min and transfer the supernatant to 25 ml volumetric flasks. Make the volumes to 25 ml with methanol. Measure the optical densities at 485 nm using the spectrophotometer.

Use of a flow cell in the spectrophotometer and an automatic sample changer allows a much larger through put of samples (80 samples/hour). The sample plate is loaded in the following order : blank (methanol), 10 μg/ml TPF standard, blank whole range of TPF standards, blank, soil extracts with blanks after every 10 extracts. The initial 10 μg/ml TPF standard is used to adjust the pen deflection on the chart recorder of the spectrophotometer.

### *(v) Calculation of results*

TPF concentrations are obtained from comparison with a standard curve of TPF in methanol. Results are recorded in volumes of hydrogen transferred during the reduction of TTC to TPF in 1 g dry soil according to the equation :—

2,3,5-triphenyltetrazoliumchloride + 2H→triphenylformazan+HCl

The formation of 1 mg of TPF requires 150.35 μl H.

$$\text{The dehydrogenase activity} = \frac{(S - B) \times F \times 25 \times 0.15035}{W}$$

Where,

S = peak height

B = baseline height

F = slope of calibration graph

W = weight of dry soil (g).

## 16. NITROGEN TRANSFORMATION : AMMONIUM, NITRITE AND NITRATE ANALYSIS

### *(i) Reagents*

*2M KCl* : 745.5 g KCl dissolve in 5 l deionized water of low conductivity. If the conductivity is higher than 30 K /cm use distilled water. Add 1 ml chloroform to inhibit nitrification in the soil extract prepared with the KCl.

*Sodium acetate buffer (pH 6)* : Dissolve 34.0 g sodium acetate trihydrate in 1 l deionized water and adjust to pH 6.0 with 2 ml 5M HCl. Add 0.02 ml Brij 35 surfactant per litre.

*Sulphalinic acid* : 5 g dissolve in 500 ml 2 M HCl.

*NED N-(1-naphthylethylenediaminedihydrochloride)* : 1 g dissolve in 500 ml deionized water. Filter and store in a dark bottle. Do not add detergent. *THIS CHEMICAL IS A SUSPECTED CARCINOGEN. DO NOT INHALE THE DUST.*

*Zinc column* : 300 x 3 mm (i.d.) glass transmission tubing packed with zinc chips.

NaOH (1g) and potassium sodium tartrate (5g) dissolve in deionized water, the solution mix and make upto 1 l with deionized water.

*Alkaline phenate solution* : 62.4 g phenol dissolve in 100 ml 27% (w/v) NaOH solution. Cool, add 20 ml acetone and make to 1 l with deionized water.

*Sodium hypochlorite* : 50 ml of a concentrated solution (containing 10% to 14% w/v available chlorine) diluted to 1 l. with deionized water.

### *(ii) Standard Solutions*

(a) $KNO_3$ (1000 μg $NO_3$ - N/ml) : 3.611 g dry at 105°C for atleast 2 h and dissolve in 500 ml deionized water. Add 1 ml chloroform and store at 4°C.

(b) $(NH_4)_2SO_4$ (1000 µg $NH_4^+$ N/ml) : 2.3475 g dry at 105□C for at least 2 h and dissolve in 500 ml deionized water. Add 1 ml chloroform and store at 4□C.

(c) Mixed standard : 10 ml solution (a) is mix with 20 ml solution (b) and make to 200 ml with 2 M KCl solution.

(d) Working standards are prepared in the range 1 µg $NH_4^+$ - N + 0.5 µg $NO_3$--N/ml to 30 ug $NH_4^+$ + 15 µg $NO_3^-$ - N/ml by diluting appropriate volumes of mixed standard (c) with 2M KCl. These working standards may be stored for not more than 7 days at 4□C.

(e) $NaNO_2$ (500 ug $NO_2^-$ - N/ml) : 1.2321 g dry at 105□C for atleast 2 h and dissolve in 500 ml deionized water. Add 1 ml chloroform and store at 4□C. 2 ml of this solution is diluted to 100 ml with 2M KCl to give a 10 µg $NO_2^-$ -N/ml standard solution. Working standards in the range 0.05 to 0.6 µg $NO_2^-$ -N/ml are prepared by diluting the 10 µg $NO_2^-$ - N/ml solution with 2M KCl.

### *(iii) Apparatus*

Auto Analyser with two channels.

Log chart recorder paper

End-over-end shaker

100 ml conical flasks

50 ml pipettes

7 cm. filter funnels

Whatman No. 1 filter paper

### *(iv) Procedure*

Samples (25 g) of treated and control soils are extracted by shaking with 50 ml of 2M KCl for 1 hour on the shaker. The extract is filtered through a Whatman No. 1 paper into 100 ml conical flasks and the filtrate analysed on the Auto Analyzer using the many folds. Nitrate and Ammonium are analyzed simultaneously at 20 samples/h on Channels 1 and 2 of the Auto-Analyzer. Sample : Wash ratio is 1:2. Nitrate samples are run through the Zn reduction column to reduce $NO_3^-$ to $NO_2^-$. The colour measured is then equivalent to $NO_2^-$ + $NO_3^-$ in the sample and $NO_3^-$ is found by subtraction. Nitrite samples are analysed at 70 samples/h on Channel 1 with a sample : Wash ratio of 2 : 1 and bypassing the Zn reduction column. Since nitrite is not normally detected this sample rate is satisfactory. If nitrite is present in measurable quantities it will be necessary to analyse at 20 samples/h with a sample : wash ratio of 1 : 2. It is necessary to use a X4 range expander on the recorder when measuring soil nitrite values.

*N.B.* All reagents should be run through the AutoAnalyzer for atleast 30 minutes before the samples are run. Standard solutions should always be run before and after a batch of samples and, if batches are large, at intervals throughout the samples.

### *(v) Calculation of results*

Draw a graph of μg N/ml against optical density using the values for the standard solutions. This should obey Beer's law at low concentration. The slope of this calibration graph can be used as a factor for calculating N concentrations in the unknowns as follows :—

$$\mu g\ N/g\ dry\ soil = S \times F \times D \times \frac{(200 + M)}{(200 - M)}$$

Where,

S = Peak height

F = Factor (Slope of calibration graph)

D = Dilution factor of sample

M = Moisture content of soil (%)

## 17. DETERMINATION OF SOIL ENZYME : PHOSPHATASE

### *(i) Reagents*

0.5% w/v disodium phenyl phosphate solution. Store at 4°C.

0.3% w/v $AlK(SO_4)_2 . 12\ H_2O$

*Sodium borate buffer, pH 9.4* : Dissolve 3.81 g disodium tetraborate decahydrate in 1 l deionized water. Adjust to pH 9.4 with M NaOH. This reagent can be stored for 1 month at 4°C.

*Gibbs reagent* : Dissolve 0.125 g 2,6-dibromoquinone-4-chloroimide in 100 ml absolute alcohol. Do not keep this reagent for more than 8 h.

*Phenol standard solutions* : A 1000 μg phenol/ml stock solution is prepared by dissolving 0.25 g phenol in 250 ml deionized water. The solution may be stored at 4°C. Working standards in the range 2.5 to 25 μg/ml are prepared from the stock solution by dilution with deionized water.

***(ii) Apparatus***

Technicon AutoAnalyzer

Log chart recorder paper

50 ml automatic dispenser

20 ml pipettes

100 ml conical flasks

7 cm filter funnels

Whatman No. 1 filter papers

***(iii) Procedure***

Samples (5 g) of treated or control soils are mixed with 20 ml disodium phenyl phosphate solution, prewarme to 30°C, in 100 ml conical flasks. The suspensions are incubated at 30°C for 2 h then mix with 100 ml $AlK(SO_4)_2$ solution and filter through Whatman No. 1 paper. The phenol content of the filtrate is measured using the AutoAnalyzer with the manifold. All reaagent should be run through the Auto Analyzer for at least 30 min before starting the analyses.

The Sample tray (8-10 ml cups) should be loaded with the standards followed by two water washes and then the soil extracts. A water wash is used after every tenth soil extract. The instrument is run at 40 samples/h with a sample wash ratio of 2:1.

***(iv) Calculations of Results***

Draw a calibration graph from the optical densities of the standards and calculate the slope of the graph.

Phosphatase activity is expressed as mg phenol released/g dry soil.

$$\text{mg phenol/g dry soil} = \frac{(S - B) \times F \times 120}{1000 \times W}$$

Where,

S = Peak height

B = Baseline height*

F = Slope of calibration graph

W = Weight of dry soil (g)

* Baseline drift occurs with this technique and must therefore be allowed for the calculation.

## 18. DETERMINATION OF SOIL ENZYME : UREASE

### *(i) Reagents*

All solutions should be made in glass-distilled water.

0.4 % w/v sodium azide. Store at 4°C.

0.5 M tris/0.5 M maleic buffer : Dissolve 6.057 g tris (hydroxy methyl) methylamine and 5.8035 g maleic acid in 100 ml water. Adjust the pH to 7.0 with M NaOH solution. This solution can be stored at 4°C for not more than 1 week.

Boric acid/ indicator solution : All glassware used in the preparation of this solution should be rinsed with ethanol. Dissolve 0.0084 g bromocresol green and 0.016 g methyl red in 10 ml ethanol. dissolve 4 g boric acid in water with heating , cool, transfer to 200 ml volumetric flask, add 4 ml indicator solution and make upto volume with water. Store at 4°C in the dark.

0.01 M $Ag_2SO_4$ : Dissolve 0.3118 g in 100 ml water with heating (avoid boiling). Store in dark at room temperature.

6.0 M urea : Make up immediately before use.

45% w/v $K_2CO_3$ 1. 1/2 $H_2O$

0.2 M & 0.02 M HCl prepared from concentrated volumetric solutions.

### *(ii) Apparatus*

Conway microdiffusion dishes

0.5, 1, 2 and 3 ml pipettes

1.0 ml self-filling burettes (0.01 ml graduations)

The Conway dishes and pipettes should be washed overnight in 2 M HCl.

### *(iii) Procedure*

Weigh samples (1g) of control or treated soil and place in the outer well of the Conway dishes. Add 0.5 ml sodium azide solution and leave for 1 h. Add 2 ml tris/meleic buffer to the soil and mix using a glass rod. Add 3 ml boric acid/indicator solution to the centre well. Care must now be taken not to move the dishes to avoid mixing the contents of outer and inner wells. Add 1 ml urea to the soil-buffer mixture in each dish. Prepare blanks by adding 1 ml distilled water in place of the urea. Incubate the dishes at room temperature (about 20□C) for 1 to 2 h, depending on the soil urease activity. Add 0.5 ml $Ag_2SO_4$ should be done in a timed sequence. Add 1 ml $K_2CO_3$ to the soil mixture, cover immediately and leave for 18 hours. Some of the soil mixture in the outer well can be removed before the dishes are moved as it must not be split into the indicator solution. Titrate the indicator solution from green back to red using first 0.2 M HCl dropwise until the indicator starts to turn purple and finally with 0.02 M HCl till the red end point is reached. Calculate the total volume of HCl taken in terms of 0.02 M HCl.

***(iv) Calculation of results***

Urease activity, as μ moles $NH_3$ evolved/g dry soil/hour

$$W = \frac{(t - b) \times 20}{W \times T}$$

Where,

t = titre, ml 0.02 M HCl

b = blank titre, ml 0.02 M HCl

W = weight dry soil (g)

T = incubation time (h)

## 19. DETERMINATION OF CELLULOSE DECOMPOSITION : SOIL BURIAL TEST

***(i) Apparatus***

Shirley Test Cloth

Disposable glass microscope slides (75 x 26 mm)

Plastic Sandwich boxes (145 x 115 x 70 mm)

Polyethylene bags (260 x 310 mm)

Desiccator

Graticule - a rigid transparent plastic strip (70 x 26 mm) marked in rectangles each of which is about 1 % of the total area.

Araldite epoxy-resin adhesive.

### *(ii) Procedure*

Mount strips of test cloth (100 x 26 mm) on glass slides fixing the ends to the backs of the slides with Exopy-resin adhesive. Araldite has been found to be suitable.

Line the plastic boxes with polyethylene and place a 10 mm layer of moist soil in the bottom. Place five cloth-covered slides on edge in this soil and fill the box with more moist soil using a glass rod to make the soil firm between the slides. This ensures good contact between soil and cloth. The boxes are each covered with polyethylene bags which are secured in place with elastic bands and then inflated using a compressed air supply. The resultant polyethylene tent over each box reduces moisture losses during incubation. The boxes are weighed and incubated at 19 + 2□C for up-to 8 weeks. Moisture losses are checked every two weeks by weighing and corrected by adding water from a hypodermic needle pushed through the polyethylene tent. At the same time the air in the tents is changed by inserting a rubber tube and flushing with compressed air. The time of incubation will vary according to the activity of the soil. If the cloth strips are incubated for too long, they will disintegrate and recovery from the soil will be impossible.

Following incubation the slides are carefully removed from the soil and soil particles gently removed from the cloth using a small artist's brush. A 70 mm long piece of cloth is removed from each slide and air-dried for 18 h at room temperature in a desiccator. The strip is weighed and the weight loss calculated as a percentage of the weight of a piece of cloth of equal size which has not been buried.

The microbial colonization of the cloth strip can be assessed crudely on the basis of the areas of cloth showing different colours (eg. areas invaded by *Fusarium* spp. or *Gliocladium roseum* are frequently pink while black areas usually denote colonization by *Stachybotrys chartarum* or *Acremonium* spp.). These coloured areas are measured by covering the cloth strip with the transparent plastic graticule and counting the number of rectangles covering each coloured area.

The graticule can also be used to give an approximate estimate of the area of cloth missing due to complete decomposition. Similarly partial decomposition can be estimated with the cloth strip held in front of a light source when small holes and areas of thinning can be easily seen.

The scoring scale used is :—

0 – no effect

13 – some thinning of the material

46 – small holes appearing

79 – large areas of cloth disintegrated

10 – complete disintegration

When required the microorganisms colonizing the cloth can be isolated for further study. Fungal hyphae and/or sporing structures can be isolated using fine glass needles while observing the cloth with a stereoscopic microscope and transferred to suitable agar medium. Other microorganisms can be isolated by removing small fibres and placing these on agar plates. Alternatively pieces of cloth can be treated in water using a Stomacher and the suspensions plated out.

# PART – IV
# METHODOLOGY FOR ECOLOGICAL STUDY AND VEGETATIONAL ANALYSIS

**PART – IV**

# METHODOLOGY FOR ECOLOGICAL STUDY AND VEGETATIONAL ANALYSIS

## 1. COMMUNITY

A Plant Community is a part of the natural vegetation, it grows, develops and attend into a functional unit and is in a state of constant flux with the environment. It has trophic organisation and metabolic pattern.

An Ecological Community can be regarded as an assemblage of species populations that has the potential of interaction. More precisely, a community may be defined as an interactive assemblage of species occurring together within a particular geographical area, a set of species whose ecological function and dynamics are in some way interdependent. These interaction include overt and critically determining compitative interaction and feeding relationships as well as more subtle manifestations, such as reliance of plant or animal for pollination and seed dispersal or of animals on plants for meeting habitat requirements.

*Physiognomy* : The form and structure for physiognomy of a Community can be described in a number of ways, each assessing somwhat different aspects, each having certain advantages and limitations, and as might be expected, each having its proponents in ecological circles.

*Growth forms* : Perhaps, the least technical characterization of a Community's structure particularly applied to plants.

***Dauserean (1951) Proposed —***

(A) Six growth forms :

Trees, Shrubs, Herbs, Bryoids, Epiphytes, Lianas

(B) Three sizes :

Tall, Medium and Low.

(C) Four functions or leaf habits :

Deciduous, Semi-deciduous, Evergreen Succulent and Evergreen Leafless.

(D) Six leaf shapes :

Needle or spine, Graminoids, Medium or Small, Broad, Compound, Thalloids.

(E) Four leaf texture :

Fibrary, Membranous, Sclerophyll, Succulent.

(F) Four aspects of coverage :

Barren or very Sparse, Discontinuous, in Tufts or Groups, Continuous.

## 2. LIFE FORM

Avarians of plant growth form reffered to as life form, was developed at the turn of the twenteeth century by Danish botanist Raunkiaer (1934).

The categories *Raunkiaer* recognised based on the amount and kind of protection afforded to the buds and shoot apices which are as follows :

(A) *Phanerophytes* : The plants are wood, shrubs and herbs surviving buds or shoot apcies born on height.

(B) *Cheamephytes* : These are small plants, herbs and creeping woody plants.

(C) *Hemicryptophyte* : Herbs growing in Rosettes and tussocks.

(D) *Geophytes* : Buds are burried in ground. Tuberous and bulbous herbs.

(E) *Hydrohalophytes* : Annual herbs.

## (3) STRATIFICATION

All the plants in the Community are not of the same size and do not occupy the same space there. The distribution of the plants in the vertical space is described as the stratification or layering.

*Stratification can be studied in aerial and under ground regions.*

## 4. ZONATION

Horizontal changes in the physical environment are reflected in zonational changes in plant and animal components of ecological communities. This is very useful in study of marine ecosystem, fresh water ecosystem and in relation to gradient.

## 5. HORIZONTAL DISPERSION

The horizontal spacing or dispersion of plants and/or animal can also be used to describe the structure of an ecological community. These are three basic patterns of dispersal :

(A) Random

(B) Uniform or Regular

(C) Clumped or Contagious

## 6. SOCIABILITY

The sociability express the relation of individuals to each other and indicates the closeness between individuals. Braun Blanquet (1932) recognised five arbitrary categories of sociability:

$S_1$ = Plants growing singly

$S_2$ = Plants growing in small groups

$S_3$ = Plants occurring in small patches

$S_4$ = Plants forming large patches

$S_5$ = Plants occurring in essentially continuous populations.

Whitford (1949) has attempted to express sociability in quantitative terms by the following equations :

Sociability = Density x 100 / Frequency

## 7. VITALITY

It is concerned with the normal growth and reproductive ability of the species which helps it in maintaining its position in community. Various authors have recognised three to five categories of Vitality :

$V_1$ = Plants which germinate but die soon without reproducing.

$V_2$ = Plants which linger after germination but can not reproduce.

$V_3$ = Plants reproducing but only vegetatively.

$V_4$ = Plants reproducing sexually but rather feebly.

***Quantitative Characters :***

**(i) Density** : Density is the number of individuals of a species in an unit area or volume.

$$\text{Density} = \frac{\text{Total number of plants in the quadrats}}{\text{Number of quadrats examined}}$$

*The Density is determined by actual count for each species in the community.*

**(ii) Frequency :** Frequency indicates the homogeniety of dispersion of the individuals of a species in the Community. The frequency is expressed as the percentage occurrence of individuals of a species in a number of observations.

$$\%\ \text{Frequency} = \frac{\text{Number of quadrats in which the species occur}}{\text{Total number of quadrats examined}} \times 100$$

Raunkiaer (1934) divided the frequency into five classes :

A = 0 - 20 %

B = 21 - 40 %

C = 41 - 60 %

D = 61 - 80 %

E = 81 - 100 %

Raunkiaer proposed a law of frequency which states that the percentage of species failing under the five classes of frequency will be in the order

$$A > B > C > D < E$$

He found that in a homogenously distributed vegetation the proportion of the species will be 53, 14, 9, 8 and 16 per cent respectively in the five classes.

**(iii) Abundance :** Abundance of a species in the Community represents its relative distribution in it. It is related to density but is a qualitative estimates.

$$\text{Abundance} = \frac{\text{Total number of plants of a species in all the quadrats}}{\text{Number of quadrat in which the species occurred}}$$

**(iv) Importance Value Index (IVI) :** Curtis and McIntosh (1950) proposed as Importance Value Index (IVI) which is sum of the relative values of the three quantitave characters. The relative value are expressed as percentage of the total values for all species in a community as given below :

*Relative Frequency* (RF)

$$= \frac{\text{Frequency of one species}}{\sum \text{Frequency of all species}} \times 100$$

*Relative Density* (RD)

$$= \frac{\text{Total number of plants of a species}}{\text{Total number of plants of all the species}} \times 100$$

*Relative Dominance* (R. Dom.)

$$= \frac{\text{Total canopy cover of species}}{\text{Total canopy cover of all the species}} \times 100$$

Importance Value Index (IVI) = RF + RD + R.Dom.

**(v) Index of Dominance**

$$C = \sum (ni/N)^2$$

Where,

ni = Importance value of each species (number, biomass, production, etc)

N = Total of all Importance value.

**(vi) Index of Similarity between two Communities**

$$IS = \frac{2\,C}{A + B}$$

Where,

A = Number of species in Community A

B = Number of species in Community B

C = Number of species common to both Communities.

IS = Index of Similarity

### (vii) Point Centered Quarter Method of Sampling Vegetation

This method has been developed by Cottan and Curtis (1949, 1956). In this experiment, four distances are measured at each sampling point. Four quarters are established at the sampling point through a cross formed by two lines. One line is the compass directed and the second line running perpendicular to the compass direction through the sampling point.

The plants nearest to the pin in the four quarters are recorded as shown in Fig. —

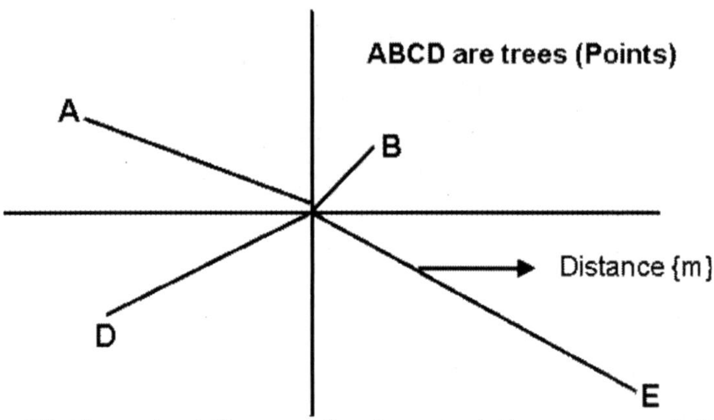


The important Community characteristics are computed as under :

**(A)** *Frequency*

$$= \frac{\text{Number of points with species present}}{\text{Total number of points observed}} \times 100$$

**(B)** *Average Distance*

$$= \frac{\text{Distance of all species recorded}}{\text{Number of points} \times 4}$$

**(C)** *Average Area* = (Average Distance)$^2$

**(D)** *Absolute Density of all the species*

$$= \frac{100 \times 100}{\text{Average Area (Cm)}^2}$$

**(E)** *Relative Density of the species*

$$= \frac{\text{Number of individuals at all points} \times 100}{\text{Number of points} \times 4}$$

**(F)** *Absolute Density of the species*

$$= \frac{\text{RD} \times \text{Absolute Density of all the species}}{100}$$

**Quantitative analysis by Point Centered Quarter Method. Five Sampling Points**

| Sampling point | Quarter number | Distance (M) | Species | Diameter at the basal area $cm^2$ |
|---|---|---|---|---|
| 1 | 1 | 0.7 | *Tectona grandis* | 5.5 |
| | 2 | 1.6 | *Acacia arabica* | 42.5 |
| | 3 | 3.5 | *Butea monosperma* | 17.0 |
| | 4 | 2.0 | *Diospyros melanoxylon* | 25.0 |
| 2 | 1 | 1.1 | *Tectona grandis* | 4.0 |
| | 2 | 0.8 | *Tectona grandis* | 5.0 |
| | 3 | 1.9 | *Tectona grandis* | 5.0 |
| | 4 | 1.8 | *Tectona grandis* | 4.0 |
| 3 | 1 | 1.3 | *Acacia arabica* | 75.0 |
| | 2 | 0.7 | *Tectona grandis* | 3.0 |
| | 3 | 1.5 | *Butea monosperma* | 9.0 |
| | 4 | 2.0 | *Butea monosperma* | 23.0 |
| 4 | 1 | 3.1 | *Acacia arabica* | 14.0 |
| | 2 | 1.7 | *Tectona grandis* | 6.0 |
| | 3 | 1.1 | *Tectona grandis* | 5.0 |
| | 4 | 1.9 | *Acacia arabica* | 12.0 |
| 5 | 1 | 2.5 | *Acacia arabica* | 23.0 |
| | 2 | 2.2 | *Acacia arabica* | 18.0 |
| | 3 | 1.4 | *Tectona grandis* | 5.0 |
| | 4 | 2.8 | *Butea monosperma* | 25.0 |

**Calculations :**

**(I) Mean Distance (D)** 35.6 / 20 = 1.78 m

**(II) Absolute Density** = Area / $D^2$

Where, D = Mean Distance

**(III) Number of trees per 100 $m^2$** $= 100 / (1.78)^2$
$= 100 / 3.17$
$= 31.5$

**(IV) Absolute Dominance**

= means 'ba' per tree x number of trees in species

Where, ba = basal area

**(V) Number of trees in species**

| | Species | Number of quater | Number of trees in 100 $M^2$ |
|---|---|---|---|
| 1 | *Acacia arabica* | 6/20 = 0.3 | 0.3 x 31.5 = 9.5 |
| 2 | *Butea monosperma* | 4/20 = 0.2 | 0.2 x 31.5 = 6.3 |
| 3 | *Diospyros melanoxylon* | 1/20 = 0.05 | 0.05 x 31.5 = 1.6 |
| 4 | *Tectona grandis* | 9/20 = 0.45 | 0.45 x 31.5 = 14.2 |
| | | | Total = 31.5 |

Mean basal area (ba) by species for the 20 lines :

*(1) Acacia arabica*

| Diameter (cm) | Basal area $(cm)^2$ |
|---|---|
| 42.5 | 1418 |
| 75.0 | 4418 |
| 14.0 | 154 |
| 12.0 | 113 |
| 23.0 | 415 |
| 18.0 | 254 |
| Total basal area | 6772 |
| Mean basal area | 1129 |

*(2) Butea monosperma*

| Diameter (cm) | Basal area $(cm)^2$ |
|---|---|
| 17.0 | 227 |
| 9.0 | 64 |
| 23.0 | 415 |
| 25.0 | 491 |
| Total basal area | 1197 |
| Mean basal area | 299 |

*(3) Diospyras melanoxylon*

| Diameter (cm) | Basal area $(cm)^2$ |
|---|---|
| 25.0 | 491 |
| Total basal area | 491 |
| Mean basal area | 491 |

**(4)** *Tectona grandis*

| Diameter (cm) | Basal area $(cm)^2$ |
|---|---|
| 5.5 | 24 |
| 4.0 | 13 |
| 5.0 | 20 |
| 5.0 | 20 |
| 4.0 | 13 |
| 3.0 | 7 |
| 6.0 | 28 |
| 5.0 | 20 |
| 5.0 | 20 |
| Total basal area | 165 |
| Mean basal area | 18 |

**(VI) Therefore Dominance of**

| | | Dominance Rank |
|---|---|---|
| *Acacia arabica* | 1129 x 9.4 = 10613 $cm^2$ | 1 |
| *Butea monosperma* | 299 x 6.3 = 1884 $cm^2$ | 2 |
| *Diospyros melanoxylon* | 491 x 1.6 = 786 $cm^2$ | 3 |
| *Tectona grandis* | 18 x 14.2 = 256 $cm^2$ | 4 |

**(VII) Absolute Frequency**

$$= \frac{\text{Number of points with species}}{\text{Total points}} \times 100$$

| | | |
|---|---|---|
| 1 | *Acacia arabica* | 4/5 x 100 = 80 Per cent |
| 2 | *Butea monosperma* | 3/5 x 100 = 60 Per cent |
| 3 | *Diospyras melanoxylon* | 1/5 x 100 = 20 Per cent |
| 4 | *Tectona grandis* | 5/5 x 100 = 100 Per cent |
| | | 260 Per cent |

## Determination of Importance Value Index

**(A) Relative Density**

$$= \frac{\text{Number of individuals of species}}{\text{Total number of individuals}} \times 100$$

| | |
|---|---|
| (1) *Acacia arabica* | 9.5/31.5 x 100 = 30 Per cent |
| (2) *Butea monosperma* | 6.3/31.5 x 100 = 20 Per cent |
| (3) *Diospyras melanoxylon* | 1.6/31.5 x 100 = 5 Per cent |
| (4) *Tectona grandis* | 14.3/31.5 x 100 = 45 Per cent |
| | = 100 per cent |

**(B) Relative Dominance**

$$= \frac{\text{Dominance of a species}}{\text{Dominance of all species}} \times 100$$

| | |
|---|---|
| (1) *Acacia arabica* | 10,613 / 13,539 x 100 = 78.4 Per cent |
| (2) *Butea monosperma* | 1,884 / 13,539 x 100 = 13.0 Per cent |
| (3) *Diospyras melanoxylon* | 786 / 13,539 x 100 = 5.8 Per cent |
| (4) *Tectona grandis* | 256 / 13,539 x 100 = 1.9 Per cent |

**(C) Relative Frequency**

$$= \frac{\text{Frequency of a species}}{\text{Frequency of all species}} \times 100$$

| | |
|---|---|
| (1) *Acacia arabica* | 80 / 260 x 100 = 30.8 Per cent |
| (2) *Butea monosperma* | 60 / 260 x 100 = 23.1 Per cent |
| (3) *Diospyras melanoxylon* | 20 / 260 x 100 = 7.7 Per cent |
| (4) *Tectona grandis* | 100 / 260 x 100 = 38.5 Per cent |
| | 100.1 Percent |

**Summarized Table**

**(D) Importance Value = RD + R.Dom. + RF**

| | Species | RD | R.Dom. | RF | IVI | IVI Rank |
|---|---|---|---|---|---|---|
| 1 | *Acacia arabica* | 30.0 | 78.4 | 30.8 | 139.2 | 1 |
| 2 | *Butea monosperma* | 20.0 | 13.9 | 23.1 | 57.0 | 3 |
| 3 | *Diospyras melanoxylon* | 5.0 | 5.8 | 7.7 | 18.5 | 4 |
| 4 | *Tectona grandis* | 45.0 | 1.9 | 38.5 | 85.4 | 2 |

**Conclusion**

From above table indicates that *Acacia arabica* and *Tectona grandis* are the dominant and co-dominant species in above Forest Community.

## 8. PRACTICAL EXERCISE RELATED TO BIOMASS, PRODUCTION AND ENERGY

**(i) Biomass** = Total weight of OM (Above ground+below ground) of particular area for given time.

$$\Delta B = B_2 - B_1$$

**(ii) Production rate** $= \dfrac{W_2 - W_1}{t_2 - t_1}$

Where,

$W_2$ is the dry weight at time $t_2$

$W_1$ is the dry weight at time $t_1$

**(iii) Turn Over rate**

$$= \frac{\text{Gross Primary Productivity (GPP)}}{\text{Standing Crop Biomass}} \text{ hr/day/year}$$

**(iv) Turn Over time** $= \dfrac{\text{Standing Crop Biomass}}{\text{GPP}}$ hr/day/year

**(v) Ecological turn over** $= \dfrac{\text{Community Respiration rate (R)}}{\text{Total Community Biomass (B)}}$

**(vi) Turn Over Value** (T) $= \dfrac{B_{max} - B_{min}}{\text{Total B}}$

Where,

$B_{max}$ is maximum biomass

$B_{min}$ is minimum biomass

Total B is total mean biomass

**(vii) % Efficiency of Energy Utilization (EU)**

$$EU = \frac{\text{Gross Production}}{\text{Total Solar Radiation}} \times 100$$

## 9. DETERMINATION OF ENERGY — VALUE OF MATERIALS BY BOMB CALORIMETER

Leith (1968) described the method of energy estimation with help of oxygen bomb calorimeter. The productivity values are then given in terms of energy. The weight measurements are not exactly comparable. Therefore, now a days, to give productivity values in terms of energy is, preferred than in terms of weight.

**Method :** 1 g of powder of dried plant materials is taken in pellet machine. The pellet is now weighed alongwith ignition wire and is placed in ignition cup in the bomb. The known quantity of water is taken in Bomb Vessel and temperature is allow to constant by the help of stirrer. When temperature becomes constant, this will be $\mathbf{t_1}$. The Calorimeter bomb filled with oxygen and placed inside the vessel before taking the reading of vessel. Now bomb is link with electrical circut for ignite the inside material. After ignition of material, there is rise in temperature of water. This is **$t_2$.** The difference of the temperature is the rise of the temperature. The energy given is equal to energy taken.

**MST = MST**

Where,

M = Mass

S = Specific heat

T = Rise in temperature

**(i) Energy (e)** = $\frac{W_1 (t_2 - t_1) + W_2 (t_2 - t_1)}{W}$ Cal/g dry wt.

Where,

$W_1$ = Water equivalant of the Cylinder (g)

$W_2$ = Values of Water in the Cylender

W = Weight of pellet (Plant part or animal part)

$t_1$ = Initial Temperature ▯C

$t_2$ = Final Temperature ▯C

**(ii) Volume (V)** = $\frac{W \Delta t - \sum C}{G}$ Cal/g dry wt.

Where,

G = Dry weight of material

$\Delta t$ = Rise in temperture

W = Water equivalant of system

$\sum C$ = Temperature rise value due to ignition.

## 10. REPRODUCTIVE CAPACITY OF PLANTS (RC)

$$RC = \frac{\text{Seed out of per plant} \times \%\ \text{Germination}}{100}$$

## 11. CONSERVATION VALUE OF PLANT (CV)

$$CV = 100\ (S_{wp} / S_{wo} \times 100)$$

Where,

CV = Conservation value of species

$S_{wp}$ = Weight of soil (dry) washed from a plot under cover of species.

$S_{wo}$ = Weight of soil (dry) washed from an identical but base plot under identical erosive factor.

## 12. DEMOGRAPHIC QUOTIENT (Cloud 1968, 1969, 1970)

$$Q = \frac{\text{Total Resource available}}{\text{Population Density} \times \text{Per Capita Consumption}}$$

Where,

Q = Quality of life

## 13. MEASUREMENT OF THE HEIGHT OF THE TREES

For measuring the heights of standing trees a number of instruments have been devised, which are variously called dendrometer, hypsometer, altimeter, clinometer, etc. They all utilize either the geometrical properties of similar triangles or the trignometrical ratios of angle.

By Sextant we measured the Angle

$$\text{Angle } \theta = \frac{\text{Perpendicular}}{\text{Base}}$$

Base = Parpendicular x Angle θ (tan θ)

$$\underset{\text{(Height of the tree)}}{\text{Parpendicular}} = \underset{\text{(Distance)}}{\text{Base}} \times \tan\theta$$

## 14. MEASUREMENT OF GIRTH OR DIAMETER OF THE TREE

$$A = \pi r^2 = \pi d^2/4 = 0.7854\ d^2$$

Where,

A = is the Area

r = is the radius

d = is the diameter

or $$A = g^2 / 4\pi = 0.07958\ g^2$$

Where,

g = is the girth.

## 15. ESTIMATION OF THE VOLUME OF LOGS AND TREES

$$V = L \times \frac{G^2}{4\pi} \times \frac{4}{144} \text{ Cu. ft.}$$

Where,

L = the length in feet of a log

G = the mid girth in inches

V = Volume

## 16. DETERMINATION OF COMPLEXITY INDEX (Holdridge, 1967)

This is a very simple methametical expression for the quantitative characters of the Community in form of Complexity Index. This cofficient was used originally only for comparing the vegetation of larger areas in relation to rainfall.

CI = number of spp. x Density x Basal area x Height x $10^{-3}$
Based on 0.1 ha.

Where,

CI = Complexity Index

## 17. CLIMATE, VEGETATION AND PRODUCTION INDEX (CVP Index)

To measure the ability of a site for plant growth in a particular climate, and to express it as an Index and than to correlate it with site class. Peterson has given the formula

$$\text{CVP Index or I} = \frac{T_v}{T_a} \times \frac{P}{12} \times \frac{GE}{100}$$

Where,

$T_v$ = the temperature of the warmest months

$T_a$ = the annual range of temperature of the warmest month

P = the annual precipitation in month

G = the length of growing season

E = the Evapo-transpiration reducer

## 18. PRINCIPLE OF COMPETITION

Gause (1934) who concluded from an experimental study on two species that two species with similar requirements can not co-exist and one of them is excluded.

The competition away the organisms may be for space, food, nutrients, water, light, gases, pollinating and dispersal agents or such other factors. Competition may be between the individuals of the same species (Intraspecific) or of different species (Interspecific).

$$\frac{dN_1}{dt} = r_1 N_1 \left(\frac{K_1 - N_1 - \alpha N_2}{K_1}\right) \text{ and}$$

$$\frac{dN_2}{dt} = r_2 N_2 \left(\frac{K_2 - N_2 - \beta N_1}{K_2}\right)$$

Where,

$\alpha$= Competative Coefficient for species 2 in relation to species 1.

$\beta$= Competative Coefficient for species 1 in relation to species 2.

K = Carying Capacity of the environment

## 19. DETERMINATION OF STOMATAL INDEX

$$SI = \frac{S}{S+E}$$

Where,

S = Number of stomata

E = Epidermal cells in a unit area

SI = Stomatal Index

Density of stomata may be calculated as number of stomata per unit area.

## 20. POPULATION ECOLOGY

### (i) Natality

It is the rate of births per unit time.

The Natality is expressed by the formula

$$B = N_n / t$$

Where,

B = births

$N_n$ = Number of new borns

t = Time

### (ii) Mortality

It is the death rate per unit time.

The mortality is expressed by the formula

$$d = D / t$$

Where,

d = death rate

D = Total number of deaths

t = time.

# PART - V

# APPENDICES

**PART - V**

# APPENDICES

## Elements

THEIR SYMBOLS, ATOMIC NUMBERS, ATOMIC WEIGHTS (1961) AND VALENCIES

| Name | Symbol | At. No. | Atomic Weight | Valency |
|---|---|---|---|---|
| Actinium | Ac | 89 | | |
| Aluminium | Al | 13 | 26.9815 | 3 |
| Antimony | Sb | 51 | 121.75 | 3,5 |
| Argon | A | 18 | 39.948 | 0 |
| Arsenic | As | 33 | 74.9216 | 3,5 |
| Barium | Ba | 56 | 137.34 | 2 |
| Beryllium | Be | 4 | 9.0133 | 2 |
| Bismuth | Bi | 83 | 208.98 | 3,5 |
| Boron | B | 5 | 10.811 | 3 |
| Bromine | Br | 35 | 79.909 | 1,3,5,7 |
| Cadmium | Cd | 48 | 112.40 | 2 |
| Caesium | Cs | 55 | 132.905 | 1 |
| Calcium | Ca | 20 | 40.08 | 2 |
| Carbon | C | 6 | 12.011 | 2,4 |
| Cerium | Ce | 58 | 140.12 | 3,4 |
| Chlorine | Cl | 17 | 35.453 | 1,3,5,7 |
| Chromium | Cr | 24 | 51.996 | 2,3,6 |
| Cobalt | Co | 27 | 58.9332 | 2,3 |
| Columbium | Cb | 41 | | 3,5 |
| Copper | Cu | 29 | 63.54 | 1,2 |
| Dysprosium | Dy | 66 | 162.50 | 3 |
| Erbium | Er | 68 | 167.26 | 3 |
| Europium | Eu | 63 | 151.96 | 2,3 |
| Fluorine | F | 9 | 18.9984 | 1 |
| Gadolinium | Gd | 64 | 157.25 | 3 |
| Gallium | Ga | 31 | 69.72 | 2,3 |
| Germanium | Ge | 32 | 72.59 | 4 |
| Gold | Au | 79 | 196.967 | 1,3 |
| Hafnium | Hf | 72 | 178.49 | 4 |

| Name | Symbol | At. No. | Atomic Weight | Valency |
|---|---|---|---|---|
| Helium | He | 2 | 4.0026 | 0 |
| Holmium | Ho | 67 | 164.93 | 3 |
| Hydrogen | H | 1 | 1.00797 | 1 |
| Illinium | Il | 61 | | |
| Indium | In | 49 | 114.82 | 3 |
| Iodine | I | 53 | 126.904 | 1,3,5,7 |
| Iridium | Ir | 77 | 192.2 | 3,4 |
| Iron | Fe | 26 | 55.847 | 2,3 |
| Krypton | Kr | 36 | 83.80 | 0 |
| Lanthanum | La | 57 | 138.91 | 3 |
| Lead | Pb | 82 | 207.19 | 2,4 |
| Lithium | Li | 3 | 6.939 | 1 |
| Lutecium | Lu | 71 | 174.97 | 3 |
| Magnesium | Mg | 12 | 24.312 | 2 |
| Manganese | Mn | 25 | 54.9380 | 2,3,4,6,7 |
| Masurium | Ma | 43 | | |
| Mercury | Hg | 80 | 200.59 | 1,2 |
| Molybdenum | Mo | 42 | 95.94 | 3,4,6 |
| Neodynium | Nd | 60 | 144.24 | 3 |
| Neon | Ne | 10 | 20.183 | 0 |
| Nickel | Ni | 28 | 58.71 | 2,3 |
| Nitrogen | N | 7 | 14.0067 | 3,5 |
| Osmium | Os | 76 | 190.2 | 2,3,4,8 |
| Oxygen | O | 8 | 15.9994 | 2 |
| Palladium | Pd | 46 | 106.4 | 2,4 |
| Phosphorus | P | 15 | 30.9738 | 3,5 |
| Platinum | Pt | 78 | 195.09 | 2,4 |
| Polonium | Po | 84 | | |
| Potassium | K | 19 | 39.102 | 1 |
| Praseodymium | Pr | 59 | 140.907 | 3 |
| Radium | Ra | 88 | 226.05 | 2 |
| Radon | Rn | 86 | 222 | 0 |
| Rhenium | Re | 75 | 186.2 | |
| Rhodium | Rh | 45 | 102.905 | 3 |
| Rubidium | Rb | 37 | 85.47 | 1 |
| Ruthenium | Ru | 44 | 101.07 | 3,6,4,8 |
| Samarium | Sm | 62 | 150.35 | 3 |

| Name | Symbol | At. No. | Atomic Weight | Valency |
|---|---|---|---|---|
| Scandium | Sc | 21 | 44.956 | 3 |
| Selenium | Se | 34 | 78.96 | 2,4,6 |
| Silicon | Si | 14 | 28.086 | 4 |
| Silver | Ag | 47 | 107.870 | 1 |
| Sodium | Na | 11 | 22.9898 | 1 |
| Strontium | Sr | 38 | 87.62 | 2 |
| Sulphur | S | 16 | 32.064 | 2,4,6 |
| Tantalum | Ta | 73 | 180.948 | 5 |
| Tellurium | Te | 52 | 127.60 | 2,4,6 |
| Terbium | Tb | 65 | 158.924 | 3 |
| Thallium | Tl | 81 | 240.37 | 1,3 |
| Thorium | Th | 90 | 232.038 | 4 |
| Thulium | Tm | 69 | 168.934 | 3 |
| Tin | Sn | 50 | 118.69 | 2,4 |
| Titanium | Ti | 22 | 47.90 | 3,4 |
| Tungsten | W | 74 | 183.85 | 6 |
| Uranium | U | 92 | 238.03 | 4,6 |
| Vanadium | V | 23 | 50.942 | 3,5 |
| Xenon | Xe | 54 | 131.3 | 0 |
| Ytterbium | Yb | 70 | 173.04 | 3 |
| Yttrium | Y | 39 | 88.905 | 3 |
| Zinc | Zn | 30 | 65.37 | 2 |
| Zirconium | Zr | 40 | 91.22 | 4 |

## Radioisotopes

| Element | Isotope | Half life | Mode of Disintegration | Maximum Energies of Radiations MeV |
|---|---|---|---|---|
| Calcium | 45 | 164 d | β – | 0.25 (100%) |
| Carbon | 14 | 5760 y | β – | 0.155 (100%) |
| Chlorine | 36 | $3.1 \times 10^5$ y | β – | 0.714 (98%) |
| Chromium | 51 | 27.8 d | E.C. | 0.323 (8%) |
| Cobalt | 60 | 5.25 y | β – | 0.31 (100%) |
| | | | γ | 1.17 (100%), 1.33 (100%) |
| Iodine | 131 | 8.0 d | β – | 0.61 (87%), 0.81 (1%) |
| | | | γ | 0.36 (80%), 0.64 (9%), 0.72 (3%) |
| Iron | 55 | 2.9 y | E.C. | |
| Iron | 59 | 45 d | β – | 0.27(46%),0.46(53%),1.56(0.3%) |
| | | | γ | 1.10 (56%), 1.29 (44%) |
| Phosphorus | 32 | 14.2 d | β – | 1.71 (100%) |
| Potassium | 42 | 12.5 h | β – | 2.0 (18%), 3.6 (82%) |
| | | | γ | 1.52 (18%) |
| Sodium | 22 | 2.6 y | β + (89%) | 0.51 (γ-radiation) |
| | | | E.C.(11%) | 1.28 (X-radiation) |
| Sodium | 24 | 15 h | β – | 1.39 (100%) |
| | | | γ – | 1.37 (100%), 2.75 (100%) |
| Sulphur | 35 | 87 d | β – | 0.167 (100%) |
| Tritium | 3 | 12.26 y | β – | 0.018 (100%) |
| Zinc | 65 | 245 d | β + (2%) | 0.51 (y -radiation) |
| | | | E.C.(98%) | 1.11 (X-radiation) |

h, d, y = hours, days, years respectively.
E.C. = Electron Capture; a radioactive transformation whereby a nucleus captures one of its orbital electrons
The mean energy of β-particles is about 1/3 of the maximum energy.

# Specific Gravity of Aqueous Solutions of $H_2SO_4$ and $HNO_3$ at 20□C

| % by weight | Specific Gravity $H_2SO_4$ | Specific Gravity $HNO_3$ | % by weight | Specific Gravity $H_2SO_4$ | Specific Gravity $HNO_3$ |
|---|---|---|---|---|---|
| 5 | 1.032 | 1.026 | 65 | 1.553 | 1.391 |
| 10 | 1.066 | 1.054 | 70 | 1.611 | 1.413 |
| 15 | 1.102 | 1.084 | 75 | 1.669 | 1.434 |
| 20 | 1.139 | 1.115 | 80 | 1.727 | 1.452 |
| 25 | 1.178 | 1.147 | 85 | 1.779 | 1.469 |
| 30 | 1.219 | 1.180 | 90 | 1.814 | 1.483 |
| 35 | 1.260 | 1.214 | 95 | 1.834 | 1.493 |
| 40 | 1.303 | 1.246 | 96 | 1.836 | 1.495 |
| 45 | 1.348 | 1.278 | 97 | 1.836 | 1.497 |
| 50 | 1.395 | 1.310 | 98 | 1.836 | 1.501 |
| 55 | 1.445 | 1.339 | 99 | 1.834 | 1.507 |
| 60 | 1.498 | 1.367 | 100 | 1.831 | 1.513 |

Linear interpolation, maximum errors :

Sulphuric acid – 0.003

Nitric acid – 0.001

## Specific Gravity of Aqueous Solutions of HCl and $NH_3$ at 20°C

| % by weight | Specific Gravity HCl | Specific Gravity $NH_3$ |
|---|---|---|
| 5 | 1.023 | 0.977 |
| 10 | 1.047 | 0.958 |
| 15 | 1.073 | 0.940 |
| 20 | 1.098 | 0.923 |
| 25 | 1.124 | 0.907 |
| 30 | 1.149 | 0.892 |
| 35 | 1.174 | — |
| 40 | 1.198 | — |

Linear interpolation, maximum errors :
HCl 0.001
$NH_3$ 0.002

## Specific Gravity of Aqueous Mixtures of Ethanol at 15.6°C

| % Alcohol by Vol. | Specific Gravity | % Alcohol by Vol. | Specific Gravity | % Alcohol by Vol. | Specific Gravity | % Alcohol by Vol. | Specific Gravity |
|---|---|---|---|---|---|---|---|
| 100 | 0.794 | 75 | 0.877 | 50 | 0.934 | 25 | 0.971 |
| 95 | 0.816 | 70 | 0.890 | 45 | 0.944 | 20 | 0.976 |
| 90 | 0.834 | 65 | 0.902 | 40 | 0.952 | 15 | 0.981 |
| 85 | 0.850 | 60 | 0.914 | 35 | 0.959 | 10 | 0.987 |
| 80 | 0.864 | 55 | 0.924 | 30 | 0.965 | 5 | 0.993 |

Linear interpolation, maximum errors : 0.001

# Viscosity

Viscosity of water at various temperatures, in Centipoises

| □C | 0 | 1 | 2 | 3 | 4 | 5 | 6 | 7 | 8 | 9 |
|---|---|---|---|---|---|---|---|---|---|---|
| 0 | 1.794 | 1.732 | 1.674 | 1.619 | 1.568 | 1.519 | 1.473 | 1.429 | 1.387 | 1.348 |
| 10 | 1.310 | 1.274 | 1.239 | 1.206 | 1.175 | 1.145 | 1.116 | 1.088 | 1.060 | 1.034 |
| 20 | 1.009 | 0.984 | 0.961 | 0.938 | 0.916 | 0.895 | 0.875 | 0.855 | 0.836 | 0.818 |
| 30 | 0.800 | 0.783 | 0.767 | 0.751 | 0.736 | 0.721 | 0.706 | 0.693 | 0.679 | 0.666 |
| 40 | 0.654 | 0.642 | 0.630 | 0.618 | 0.608 | 0.597 | 0.587 | 0.577 | 0.568 | 0.558 |
| 50 | 0.549 | 0.541 | 0.532 | 0.534 | 0.515 | 0.507 | 0.499 | 0.492 | 0.484 | 0.477 |
| 60 | 0.470 | 0.463 | 0.456 | 0.450 | 0.443 | 0.437 | 0.431 | 0.425 | 0.419 | 0.413 |

| 70□ | 80□ | 90□ | 100□ |
|---|---|---|---|
| 0.407 | 0.357 | 0.317 | 0.284 |

Viscosity of various liquids at 20□C in Centipoises

| | | | |
|---|---|---|---|
| Acetic acid | 1.3 | Ethanol | 1.2 |
| Acetone | 0.33 | Ethylene glycol | 19.9 |
| Benzene | 0.65 | Glycerol | 1500 |
| Carbon tetrachloride | 1.0 | Mercury | 1.55 |
| Chloroform | 0.56 | Sulphuric acid | 26 |
| Ether | 0.23 | Turpentine | 1.5 |

## Surface Tension of Water

Dynes per cm

| Temperature °C | Surface tension | Temperature °C | Surface tension | Temperature °C | Surface tension |
|---|---|---|---|---|---|
| 0 | 75.6 | 25 | 72.0 | 50 | 67.9 |
| 5 | 74.9 | 30 | 71.2 | 60 | 66.2 |
| 10 | 74.2 | 35 | 70.4 | 70 | 64.4 |
| 15 | 73.5 | 40 | 69.6 | 80 | 62.6 |
| 20 | 72.8 | 45 | 68.7 | 100 | 58.9 |

## Osmotic Pressure

In a dilute solution, let :

the volume of the solvent be — v ml,
the weight of substance dissolved be — *g* grams
the molecular weight of the substance be — *m*
and the temperature be — *t*°C

then the osmotic pressure of the solution, *p*, (in mm Hg) is given by

$$p = \frac{g \times (273 + t) \times 760 \times 22{,}400}{273\, mv}$$

## Vapour Pressure of Ice and Water

In mm of mercury

| | | 0 | +1 | +2 | +3 | +4 | +5 | +6 | +7 | +8 | +9 |
|---|---|---|---|---|---|---|---|---|---|---|---|
| Ice | -10 | 1.950 | 2.131 | 2.326 | 2.537 | 2.765 | 3.013 | 3.280 | 3.568 | 3.880 | 4.217 |
| Water | 0 | 4.579 | 4.926 | 5.294 | 5.685 | 6.101 | 6.543 | 7.013 | 7.513 | 8.045 | 8.609 |
| | 10 | 9.209 | 9.844 | 10.518 | 11.231 | 11.987 | 12.788 | 13.634 | 14.530 | 15.477 | 16.477 |
| | 20 | 17.535 | 18.650 | 19.827 | 21.068 | 22.377 | 23.756 | 25.209 | 26.739 | 28.349 | 30.043 |
| | 30 | 31.824 | 33.695 | 35.663 | 37.729 | 39.898 | 42.175 | 44.563 | 47.067 | 46.692 | 52.442 |
| | 40 | 55.324 | 58.34 | 61.50 | 64.80 | 68.26 | 71.88 | 75.65 | 79.60 | 83.71 | 88.02 |
| | 50 | 92.51 | 97.20 | 102.09 | 107.20 | 112.51 | 118.04 | 123.80 | 129.82 | 136.08 | 142.60 |
| | 60 | 149.38 | 156.43 | 163.77 | 171.38 | 179.31 | 187.54 | 196.09 | 204.96 | 214.17 | 223.73 |
| | 70 | 233.7 | 243.9 | 254.6 | 265.7 | 277.2 | 289.1 | 301.4 | 314.1 | 327.3 | 341.0 |
| | 80 | 355.1 | 369.7 | 384.9 | 400.6 | 416.8 | 433.6 | 450.9 | 468.7 | 487.1 | 506.1 |
| | 90 | 525.76 | 546.05 | 566.99 | 588.60 | 610.90 | 633.90 | 657.62 | 682.07 | 707.27 | 733.24 |
| | 100 | 760.00 | 787.57 | | | | | | | | |

Linear interpolation (to 0.1□C) introduces maximum errors of ± 0.05 per cent

## Wet and Dry Bulb Hygrometer (Ventilated type)

Relative humidities

| Depression of Wet Bulb °C | Dry Bulb Temperature °C | | | | | | | | |
|---|---|---|---|---|---|---|---|---|---|
| | 0 | 5 | 10 | 15 | 20 | 25 | 30 | 35 | 40 |
| 1 | 81 | 87 | 88 | 89 | 90 | 92 | 93 | 93 | 94 |
| 2 | 64 | 72 | 76 | 80 | 82 | 85 | 86 | 87 | 88 |
| 3 | 46 | 59 | 66 | 71 | 74 | 77 | 79 | 81 | 82 |
| 4 | 29 | 45 | 55 | 62 | 66 | 70 | 73 | 75 | 76 |
| 5 | 13 | 33 | 44 | 53 | 59 | 63 | 67 | 70 | 72 |
| 6 | – | 21 | 34 | 44 | 52 | 57 | 61 | 64 | 66 |
| 7 | – | 9 | 25 | 36 | 45 | 50 | 55 | 59 | 61 |
| 8 | – | – | 15 | 28 | 38 | 44 | 50 | 54 | 56 |
| 9 | – | – | 6 | 20 | 30 | 38 | 44 | 50 | 52 |
| 10 | – | – | – | 13 | 24 | 33 | 39 | 44 | 48 |

More complete data can be found in *Hygrometric Tables* of the Meteorological Office (Air Ministry) published by H.M.S.O.

## Concentration of Commerical Acids and Ammonia (approx.)

| Acids | % by weight | Specific gravity | Grams per litre | Mols. per litre | Normal solution ml. of conc. reagent per litre |
|---|---|---|---|---|---|
| Sulphuric acid | 98 | 1.835 | 1800 | 18.3 | 27.3 |
| Nitric acid | 71 | 1.420 | 1006 | 16.0 | 62.5 |
| Hydrochloric acid | 36 | 1.18 | 424 | 11.6 | 86.1 |
| Ammonia | 35 | 0.88 | 270 | 15.9 | 63.0 |

## pH of Aqueous Solutions of Acids and Bases (approx.)

| Acids | Concen-tration | pH | Base | Concen-tration | pH |
|---|---|---|---|---|---|
| Hydrochloric | N | 0.1 | Sodium hydroxide | N | 14.0 |
| Hydrochloric | 0.1N | 1.1 | Sodium hydroxide | 0.1N | 13.0 |
| Hydrochloric | 0.01N | 2.0 | Sodium hydroxide | 0.01N | 12.0 |
| Sulphuric | N | 0.3 | Potassium hydroxide | N | 14.0 |
| Sulphuric | 0.1N | 1.2 | Potassium hydroxide | 0.1N | 13.0 |
| Sulphuric | 0.01N | 2.1 | Potassium hydroxide | 0.01N | 12.0 |
| Sulphurous | 0.1N | 1.5 | Lime | Satur-ated | 12.4 |
| Orthophosphoric | 0.1N | 1.5 | Trisodium phosphte | 0.1N | 12.0 |
| Oxalic | 0.1N | 1.6 | Sodium carbonate | 0.1N | 11.6 |
| Tartaric | 0.1N | 2.2 | Ammonia | N | 11.6 |
| Malic | 0.1N | 2.2 | Ammonia | 0.1N | 11.1 |
| Citric | 0.1N | 2.2 | Ammonia | 0.01N | 10.6 |
| Formic | 0.1N | 2.3 | Potassium cyanide | 0.1N | 11.0 |
| Lactic | 0.1N | 2.4 | Magnesia | Satur-ated | 10.5 |
| Acetic | N | 2.4 | Calcium carbonate | Satur-ated | 9.4 |
| Acetic | 0.1N | 2.9 | Borax | 0.1N | 9.2 |
| Acetic | 0.01N | 3.4 | Sodium bicarbonate | 0.1N | 8.4 |
| Cabonic | Satur-ated | 3.8 | | | |
| Hydrocyanic | 0.1N | 5.1 | | | |
| Boric | 0.1N | 5.2 | | | |

## pK of Acids and Bases in Aqueous Solution

| | | | |
|---|---|---|---|
| Trichloracetic Acid | 0.70 | Propionic Acid | 4.87 |
| Iodic Acid | 0.72 | Malic Acid (2) | 5.02 |
| Picric Acid | 0.80 | Pyridine | 5.19 |
| Oxalic Acid (1) | 1.19 | Citric Acid (3) | 5.40 |
| L-Histidine (1) | 1.77 | Phthalic Acid (2) | 5.51 |
| Maleic Acid (1) | 1.82 | Succinic Acid (2) | 5.55 |
| Phosphoric Acid (1) | 1.96 | Uric Acid | 5.82 |
| L-Aspartic Acid (1) | 2.10 | L-Histidine (2) | 6.10 |
| L-Glutamic Acid (1) | 2.10 | Carbonic Acid (1) | 6.37 |
| L-Lysine (1) | 2.18 | Maleic Acid (2) | 6.58 |
| Glycine (1) | 2.35 | Phosphoric Acid (2) | 7.12 |
| Monochloracetic Acid | 2.85 | Glycylglycine (2) | 8.07 |
| Phthalic Acid (1) | 2.90 | Tris (Hydroxymethyl) aminomethane | 8.14 |
| Tartaric Acid (1) | 2.96 | L-Lysine | 8.95 |
| Fumaric Acid (1) | 3.0 | L-Histidine (3) | 9.13 |
| Citric Acid (1) | 3.08 | Boric Acid | 9.19 |
| Glycylglycine (1) | 3.14 | L-Glutamic Acid (3) | 9.47 |
| Malic Acid (1) | 3.4 | Glycine (2) | 9.78 |
| β - alanine (1) | 3.6 | Trimethylamine | 9.80 |
| Formic Acid | 3.76 | L-Aspartic Acid (3) | 9.82 |
| L-Aspartic Acid (2) | 3.86 | Phenol | 9.89 |
| Lactic Acid | 3.86 | Hydroquinone | 9.96 |
| Barbituric Acid | 3.98 | β - Alanine (2) | 10.19 |
| α- naphthylamine | 3.92 | Carbonic Acid (2) | 10.36 |
| L-Glutamic Acid (2) | 4.07 | L-Lysine (3) | 10.53 |
| β - naphthylamine | 4.11 | Methylamine | 10.64 |
| Tartaric Acid (2) | 4.16 | Ethylamine | 10.67 |
| Succinic Acid (1) | 4.18 | Dimethylamine | 10.70 |
| Oxalic Acid (2) | 4.21 | Trimethylamine | 10.72 |
| Phenylacetic Acid | 4.31 | Piperidine | 11.12 |
| Fumaric Acid (2) | 4.52 | Diethylamine | 12.00 |
| Aniline | 4.58 | Phosphoric Acid (3) | 12.32 |
| Citric Acid (2) | 4.75 | Salicylic Acid | 13.00 |
| Acetic Acid | 4.76 | | |
| Benzoic Acid | 4.80 | | |

$pK = \log_{10} \frac{1}{K}$, where K = dissociation constant.

The figure in parentheses refers to the hydrogen concerned.

# Buffers

ACETATE : HCl (WALPOLE) pH 0.65 to 5.20
*N* Sodium acetate
*N* HCl

Add 50 ml of *N* sodium acetate to amount of *N* HCl shown in table, and make up to 250 ml

| pH | ml HCl | pH | ml HCl | pH | ml HCl | pH | ml HCl |
|---|---|---|---|---|---|---|---|
| 0.65 | 100 | 1.71 | 55 | 3.09 | 48.5 | 4.19 | 35.0 |
| 0.75 | 90 | 1.85 | 53.5 | 3.29 | 47.5 | 4.39 | 30.0 |
| 0.91 | 80 | 1.99 | 52.5 | 3.49 | 46.25 | 4.58 | 25.0 |
| 1.09 | 70 | 2.32 | 51.0 | 3.61 | 45.0 | 4.76 | 20.0 |
| 1.24 | 65 | 2.64 | 50.0 | 3.79 | 42.5 | 4.92 | 15.0 |
| 1.42 | 60 | 2.72 | 49.75 | 3.95 | 40.0 | 5.20 | 10.0 |

ACETATE : ACETIC ACID (WALPOLE) pH 3.6 to 5.6
0.1 *N* Sodium acetate
0-1 *N* Acetic acid

| pH | 0.1*N* sodium acetate (ml) | 0.1*N* acetic acid (ml) | pH | 0.1*N* sodium acetate (ml) | 0.1*N* acetic acid (ml) |
|---|---|---|---|---|---|
| 3.6 | 15 | 185 | 4.8 | 120 | 80 |
| 3.8 | 24 | 176 | 5.0 | 141 | 59 |
| 4.0 | 36 | 164 | 5.2 | 158 | 42 |
| 4.2 | 53 | 147 | 5.4 | 171 | 29 |
| 4.4 | 74 | 126 | 5.6 | 181 | 19 |
| 4.6 | 98 | 102 | | | |

PHTHALATE : HCl (CLARK & LUBS) pH 2.2 to 3.8
0.2*M* Potassium acid phthalate
0.2 *M* HCl

To 50 ml of the 0.2M phthalate add $x$ ml of 0.2M HCl and dilute to 200 ml

| pH | ml 0.2 *M* HCl ($x$) | pH | ml 0.2 *M* HCl ($x$) | pH | ml 0.2 *M* HCl ($x$) |
|---|---|---|---|---|---|
| 2.2 | 46.60 | 2.8 | 26.50 | 3.4 | 9.95 |
| 2.4 | 39.60 | 3.0 | 20.40 | 3.6 | 6.00 |
| 2.6 | 33.00 | 3.2 | 14.80 | 3.8 | 2.65 |

PHTHALATE : NaOH (CLARK & LUBS) pH 4.0 to 6.2
0.2*M* Potassium acid phthalate
0.2 *M* NaOH

To 50 ml of the 0.2*M* phthalate add *x* ml of 0.2*M* NaOH and dilute to 200 ml

| pH | ml 0.2 *M* NaOH (*x*) | pH | ml 0.2 *M* NaOH (*x*) | pH | ml 0.2 *M* NaOH (*x*) |
|---|---|---|---|---|---|
| 4.0 | 0.40 | 4.8 | 17.50 | 5.6 | 39.70 |
| 4.2 | 3.65 | 5.0 | 23.65 | 5.8 | 43.10 |
| 4.4 | 7.35 | 5.2 | 29.75 | 6.0 | 45.40 |
| 4.6 | 12.00 | 5.4 | 35.25 | 6.2 | 47.00 |

KCl : HCl (CLARK & LUBS) pH 1.2 to 2.2
0.2 *M* KCl
0.2 *M* HCl

To 50 ml of 0.2 *M* KCl add *x* ml 0.2 *M* HCl and dilute to 200 ml.

| pH | ml 0.2*M* HCl (*x*) | pH | ml 0.2*M* HCl (*x*) |
|---|---|---|---|
| 1.2 | 64.5 | 1.8 | 16.6 |
| 1.4 | 41.5 | 2.0 | 10.6 |
| 1.6 | 26.3 | 2.2 | 6.7 |

CITRATE : HCl (SORENSEN) pH 1.04 to 4.96
21.008 g citric acid. $H_2O$ + 200 ml *N* NaOH per litre
0.1 *N* HCl
To *x* ml of the citrate, add (10 – *x*) ml of 0.1 *N* HCl

| pH | ml citrate (*x*) | pH | ml citrate (*x*) | pH | ml citrate (*x*) | pH | ml citrate (*x*) |
|---|---|---|---|---|---|---|---|
| 1.04 | 0.0 | 2.27 | 3.33 | 3.69 | 5.0 | 4.65 | 8.0 |
| 1.17 | 1.0 | 2.97 | 4.0 | 3.95 | 5.5 | 4.83 | 9.0 |
| 1.42 | 2.0 | 3.36 | 4.5 | 4.16 | 6.0 | 4.89 | 9.5 |
| 1.93 | 3.0 | 3.53 | 4.75 | 4.45 | 7.0 | 4.96 | 10.0 |

CITRATE : NaOH (SORENSEN, WALBUM'S VALUES) pH 4.96 to 6.69 20□C

21.008 g Citric acid. $H_2O$ + 200 ml *N* NaOH per litre

0.1 N NaOH

To *x* ml of the citrate add (10 – *x*) ml of 0.1 *N* NaOH.

| pH | ml citrate (*x*) | pH | ml citrate (*x*) |
|---|---|---|---|
| 4.96 | 10.0 | 5.57 | 7.0 |
| 5.02 | 9.5 | 5.98 | 6.0 |
| 5.11 | 9.0 | 6.34 | 5.5 |
| 5.31 | 8.0 | 6.69 | 5.25 |

CITRICACID : $Na_2HPO_4$ (MCILVAINE) pH 2.2 to 8.0

0.1 *M* Citric acid

0.2 *M* $Na_2HPO_4$

To *x* ml of 0.2 *M* $Na_2HPO_4$ add (20 – *x*) ml 0.1 *M* Citric acid.

| pH | 0.2 *M* $Na_2HPO_4$ ml (*x*) | pH | 0.2 *M* $Na_2HPO_4$ ml (*x*) | pH | 0.2 *M* $Na_2HPO_4$ ml (*x*) | pH | 0.2 *M* $Na_2HPO_4$ ml (*x*) | pH | 0.2 *M* $Na_2HPO_4$ ml (*x*) |
|---|---|---|---|---|---|---|---|---|---|
| 2.2 | 0.40 | 3.4 | 5.70 | 4.6 | 9.35 | 5.8 | 12.09 | 7.0 | 16.47 |
| 2.4 | 1.24 | 3.6 | 6.44 | 4.8 | 9.86 | 6.0 | 12.63 | 7.2 | 17.39 |
| 2.6 | 2.18 | 3.8 | 7.10 | 5.0 | 10.30 | 6.2 | 13.22 | 7.4 | 18.17 |
| 2.8 | 3.17 | 4.0 | 7.71 | 5.2 | 10.72 | 6.4 | 13.85 | 7.6 | 18.73 |
| 3.0 | 4.11 | 4.2 | 8.28 | 5.4 | 11.15 | 6.6 | 14.55 | 7.8 | 19.15 |
| 3.2 | 4.94 | 4.4 | 8.82 | 5.6 | 11.60 | 6.8 | 15.45 | 8.0 | 19.45 |

$Na_2HPO_4$ : $KH_2PO_4$ (SORENSEN) pH 5.29 to 8.04 18□C

11.876 g $Na_2HPO_4 . 2H_2O$ per litre

9.078 g $KH_2PO_4$ per litre

To *x* ml of the $Na_2HPO_4 . 2H_2O$ solution add (10 - *x*) ml of the $KH_2PO_4$ solution.

| pH | $Na_2HPO_4$ sol. ml (*x*) | pH | $Na_2HPO_4$ sol. ml (*x*) | pH | $Na_2HPO_4$ sol. ml (*x*) |
|---|---|---|---|---|---|
| 5.29 | 0.25 | 6.47 | 3.0 | 7.17 | 70 |
| 5.59 | 0.5 | 6.64 | 4.0 | 7.38 | 80 |
| 5.91 | 1.0 | 6.81 | 5.0 | 7.73 | 9.0 |
| 6.24 | 2.0 | 6.98 | 6.0 | 8.04 | 9.5 |

GLYCINE : HCl (SORENSEN) pH 1.04 to 3.68 18°C
Glycine and NaCl each at 0.1 *M* strength
0.1 *N* HCl

To $x$ ml of 0.1 *M* glycine-sodium chloride add $(10 - x)$ ml 0.1 *N* HCl

| pH | 0.1 *M* Glycine ml ($x$) | pH | 0.1 *M* Glycine ml ($x$) | pH | 0.1 *M* Glycine ml ($x$) |
|---|---|---|---|---|---|
| 1.04 | 0.0 | 1.65 | 4.0 | 2.92 | 8.0 |
| 1.15 | 1.0 | 1.93 | 5.0 | 3.34 | 9.0 |
| 1.25 | 2.0 | 2.28 | 6.0 | 3.68 | 9.5 |
| 1.42 | 3.0 | 2.61 | 7.0 | | |

GLYCINE : NaOH (SORENSEN, WALBUM'S VALUES) pH 8.53 to 12.90, 20°C
7.505 g Glycine + 5.85 g NaCl per litre
0.1 *N* NaOH

To $x$ ml of glycine NaCl solution add $(10 - x)$ ml 0.1 *N* NaOH

| pH | Glycine NaCl ml ($x$) | pH | Glycine NaCl ml ($x$) | pH | Glycine NaCl ml ($x$) |
|---|---|---|---|---|---|
| 8.53 | 9.5 | 10.42 | 5.5 | 12.33 | 4.0 |
| 8.88 | 9.0 | 11.01 | 5.1 | 12.60 | 3.0 |
| 9.31 | 8.0 | 11.25 | 5.0 | 12.79 | 2.0 |
| 9.66 | 7.0 | 11.51 | 4.9 | 12.90 | 1.0 |
| 10.09 | 6.0 | 12.04 | 4.5 | | |

BORATE : HCl (SORENSEN, WALBUM'S VALUES) pH 9.23 to 7.61, 20°C
12.404 g $H_3BO_3$ + 100 ml *N* NaOH per litre
0.1 *N* HCl

To $x$ ml of borate add $(10 - x)$ ml 0.1 *N* HCl

| pH | ml borate ($x$) | pH | ml borate ($x$) | pH | ml borate ($x$) |
|---|---|---|---|---|---|
| 9.23 | 10.0 | 8.89 | 8.0 | 8.27 | 6.0 |
| 9.15 | 9.5 | 8.79 | 7.5 | 8.13 | 5.75 |
| 9.07 | 9.0 | 8.67 | 7.0 | 7.93 | 5.5 |
| 8.99 | 8.5 | 8.49 | 6.5 | 7.61 | 5.25 |

BORATE : NaOH (SORENSEN, WALBUM'S VALUES) pH 9.24 to 12.38, 18°C
12.404 g $H_3BO_3$ + 100 ml *N* NaOH per litre
0.1 *N* NaOH

To *x* ml of borate add (10 – *x*) ml 0.1 *N* NaOH

| pH | ml borate (*x*) | pH | ml borate (*x*) |
|---|---|---|---|
| 9.24 | 10 | 9.97 | 6 |
| 9.36 | 9 | 11.08 | 5 |
| 9.50 | 8 | 12.38 | 4 |
| 9.68 | 7 | | |

BORIC ACID : NaOH (CLARK & LUBS) pH 7.8 to 10.0
Boric acid and potassium chloride each at 0.2 *M* strength
0.2 *M* NaOH
To 50 ml of the 0.2 *M* boric acid-potassium chloride add *x* ml of 0.2 *M* NaOH and dilute to 200 ml

| pH | ml 0.2*M* NaOH (*x*) | pH | ml 0.2*M* NaOH (*x*) | pH | ml 0.2*M* NaOH (*x*) |
|---|---|---|---|---|---|
| 7.8 | 2.65 | 8.6 | 12.00 | 9.4 | 32.00 |
| 8.0 | 4.00 | 8.8 | 16.40 | 9.6 | 36.85 |
| 8.2 | 5.90 | 9.0 | 21.40 | 9.8 | 40.80 |
| 8.4 | 8.55 | 9.2 | 26.70 | 10.0 | 43.90 |

TRIS : HCl (GOMORI) pH 7.19 to 9.10
0.2 *M* Tris (hydroxymethyl) aminomethane
0.1*N* HCl
To 25 ml 0.2 *M* Tris add *x* ml 0.1 *N* HCl, and dilute to 100 ml

| pH | ml 0.1 *N* HCl (*x*) | pH | ml 0.1 *N* HCl (*x*) | pH | ml 0.1 *N* HCl (*x*) |
|---|---|---|---|---|---|
| 7.19 | 45.0 | 7.96 | 30.0 | 8.51 | 15.0 |
| 7.36 | 42.5 | 8.05 | 27.5 | 8.62 | 12.5 |
| 7.54 | 40.0 | 8.14 | 25.0 | 8.74 | 10.0 |
| 7.66 | 37.5 | 8.23 | 22.5 | 8.92 | 7.5 |
| 7.77 | 35.0 | 8.32 | 20.0 | 9.10 | 5.0 |
| 7.87 | 32.5 | 8.41 | 17.5 | | |

VERONAL : HCl (MICHAELIS) pH 6.80 to 9.60

0.1 *M* Sodium diethylbarbiturate (20-60 g per litre)

0.1 *M* HCl

To *x* ml of 0.1 *M* barbitone add (10 - *x*) ml 0.1 *M* HCl

| pH | 0.1 *N* barbitone ml (*x*) | pH | 0.1 *N* barbitone ml (*x*) | pH | 0.1 *N* barbitone ml (*x*) |
|---|---|---|---|---|---|
| 6.80 | 5.22 | 7.80 | 6.62 | 8.80 | 9.08 |
| 7.00 | 5.36 | 8.00 | 7.16 | 9.00 | 9.36 |
| 7.20 | 5.54 | 8.20 | 7.69 | 9.20 | 9.52 |
| 7.40 | 5.81 | 8.40 | 8.23 | 9.40 | 9.74 |
| 7.60 | 6.15 | 8.60 | 8.71 | 9.60 | 9.85 |

VERONAL : HCl (MICHAELIS) pH 2.62 to 9.16

9.714 g Sodium acetate. $3H_2O$ + 14.714 g sodium diethylbarbiturate dissolved in water and made up to 500 ml

8.5 per cent NaCl (may be omitted)

0.1 *N* HCl

To 5 ml of the veronal-acetate solution add 2 ml of the NaCl solution, *x* ml of 0.1 *N* HCl and (18 – *x*) ml of water

| pH | ml 0.1 *N* HCl (*x*) | pH | ml 0.1 *N* HCl (*x*) | pH | ml 0.1 *N* HCl (*x*) |
|---|---|---|---|---|---|
| 2.62 | 16.0 | 4.43 | 9.0 | 7.66 | 4.0 |
| 3.20 | 15.0 | 5.32 | 8.0 | 7.90 | 3.0 |
| 3.62 | 14.0 | 6.12 | 7.0 | 8.18 | 2.0 |
| 3.88 | 13.0 | 6.75 | 6.5 | 8.55 | 1.0 |
| 4.13 | 12.0 | 6.99 | 6.0 | 8.68 | 0.75 |
| 4.33 | 11.0 | 7.25 | 5.5 | 8.85 | 0.5 |
| 4.66 | 10.0 | 7.42 | 5.0 | 9.16 | 0.25 |

## SALINE SOLUTIONS
### Weight in grams per litre of solution

| Fluid | NaCl | KCl | $CaCl_2$ | $NaHCO_3$ | $Na_2HPO_4$ | $NaH_2PO_4$ | $MgCl_2$ | Glucose | Remarks |
|---|---|---|---|---|---|---|---|---|---|
| Invertebrates (Hedon-Fleig) | 7.0 | 0.3 | 0.1 | 1.5 | | | $MgSO_4$ 0.3 | | |
| Insect saline | 7.5 | | | | | | | | |
| Insect saline (Clarke) | 6.5 | 1.4 | 0.12 | 0.1 | 0.01 | | | | |
| Crustacea, marine (Gatenby) | 29.23 | 0.75 | 4.44 | | | | | | |
| Mollusca, marine (Gatenby) | 23.38 | | 5.55 | | | | 7.62 | | |
| Elasmobranch (Nichol) | 16.38 | 0.89 | 1.11 | 0.38 | | 0.06 | | | Urea 21.6, Dextrose 1.0 |
| Selachian (Young) | 22.00 | 0.52 | 0.44 | | | | 0.47 | | Urea 29.0 |
| Teleost, marine (Young) | 13.50 | 0.60 | 0.25 | | | | 0.35 | | |
| Teleost, fresh-water (Young) | 5.50 | 0.14 | 0.12 | | | | | | Also for Lamprey |
| Amphibia saline | 6.00 | | | | | | | | |
| "" (Ringer) | 6.00 | 0.075 | 0.10 | 0.10 | | | | | |
| "" (Ringer, Rough) | 6.60 | 0.015 | 0.015 | to pH 7.8 | | | | | |
| "" (Ringer, Clarke) | 6.50 | 0.14 | 0.12 | 0.10 | 0.01 | | | 2.0 | |

| Fluid | NaCl | KCl | $CaCl_2$ | $NaHCO_3$ | $Na_2HPO_4$ | $NaH_2PO_4$ | $MgCl_2$ | Glucose | Remarks |
|---|---|---|---|---|---|---|---|---|---|
| "" "" (Holtfreter) | 3.50 | 0.05 | 0.10 | 0.20 | | | | | |
| "" "" (Flickinger) A | 6.80 | 0.10 | $Ca(NO_3)_2$ $4H_2O$ 0.16 | | | | $MgSO_4$0.20 | | Mix A:B:C = 2:1:1 |
| B | | | | | 0.44 | $KH_2PO_4$0.8 | | | pH = 7.8 |
| C | | | | 0.80 | | | | | |
| Chick embryo (Ringer) | 9.0 | 0.42 | 0.24 | | | | | | |
| "" "" "" buffered | 8.6 | 0.31 | 0.20 | | $K_2HPO_4$0.8 | $KH_2PO_4$0.2 | 0.10 | 2.0 | |
| "" "" "" (Locke) | 9.0 | 0.4 | 0.24 | 0.2 | | | | | |
| Mammal saline | 9.0 | | | | | | | | |
| "" "" (Ringer, Locke) | 9.2 | 0.4 | 0.24 | 0.15 | | | | For perfusion 1.0 | |
| "" "" (Ringer, Dale) | 9.0 | 0.42 | 0.24 | 0.5 | | | | | |
| Tyrode | 8.0 | 0.20 | 0.20 | 0.1 | | 0.05 | 0.20 | 1.0 | |
| Pannett & Compton A | 121.1 | 15.5 | 7.7 | | | | 0.4 | | |
| B | | | | | 1.887 | 0.145 | | | A4 ml, B 6 ml |
| Plant material (Ringer) | 7.5 | 0.075 | 0.1 | 0.1 | | | | | Distilled water 90 ml |

## Ionic Concentrations in Sea-Water
## Grams per litre (20□C) of sea-water for chlorinity 1-23% (*S* = 1.84 - 41.55 %)

| Cl % | S % | Cl | Br | $SO_4$ | $HCO_3$ | Ca | Mg | K | Sr | Na | $H_3BO_3$ |
|---|---|---|---|---|---|---|---|---|---|---|---|
| 1 | 1.84 | 0.9986 | 0.003399 | 0.1452 | 0.05404 | 0.03564 | 0.06882 | 0.01999 | 0.0007018 | 0.5554 | 0.001370 |
| 2 | 3.64 | 2.0000 | 0.006807 | 0.2846 | 0.05887 | 0.05596 | 0.1358 | 0.04004 | 0.001405 | 1.1124 | 0.002743 |
| 3 | 5.45 | 3.0040 | 0.01022 | 0.4244 | 0.06372 | 0.07633 | 0.2030 | 0.06014 | 0.002111 | 1.6708 | 0.004120 |
| 4 | 7.25 | 4.0108 | 0.01365 | 0.5646 | 0.06858 | 0.09676 | 0.2704 | 0.08030 | 0.002819 | 2.2308 | 0.005501 |
| 5 | 9.06 | 5.0204 | 0.01709 | 0.7052 | 0.07345 | 0.1172 | 0.3379 | 0.1005 | 0.003528 | 2.7923 | 0.006885 |
| 6 | 10.86 | 6.0327 | 0.02053 | 0.8462 | 0.07834 | 0.1378 | 0.4057 | 0.1208 | 0.004239 | 3.3553 | 0.008274 |
| 7 | 12.67 | 7.0476 | 0.02399 | 0.9875 | 0.08323 | 0.1584 | 0.4736 | 0.1411 | 0.004953 | 3.9198 | 0.009665 |
| 8 | 14.47 | 8.0654 | 0.02745 | 1.1292 | 0.08815 | 0.1790 | 0.5417 | 0.1615 | 0.005668 | 4.4859 | 0.01106 |
| 9 | 16.28 | 9.0858 | 0.03092 | 1.2713 | 0.09307 | 0.1997 | 0.6100 | 0.1819 | 0.006385 | 5.0534 | 0.01246 |
| 10 | 18.08 | 10.1089 | 0.03441 | 1.4138 | 0.09801 | 0.2205 | 0.6784 | 0.2024 | 0.007104 | 5.6225 | 0.01386 |
| 11 | 19.89 | 11.1348 | 0.03790 | 1.5567 | 0.1030 | 0.2413 | 0.7471 | 0.2229 | 0.007825 | 6.1930 | 0.01527 |
| 12 | 21.69 | 12.1634 | 0.04140 | 1.6999 | 0.1079 | 0.2622 | 0.8159 | 0.2435 | 0.008548 | 6.7651 | 0.01668 |

| Cl % | S % | Cl | Br | $SO_4$ | $HCO_3$ | Ca | Mg | K | Sr | Na | $H_3BO_3$ |
|---|---|---|---|---|---|---|---|---|---|---|---|
| 13 | 23.50 | 13.1947 | 0.04491 | 1.8435 | 0.1129 | 0.2831 | 0.8849 | 0.2642 | 0.009273 | 7.3388 | 0.01810 |
| 14 | 25.30 | 14.2288 | 0.04843 | 1.9875 | 0.1179 | 0.3041 | 0.9541 | 0.2849 | 0.009999 | 7.9139 | 0.01951 |
| 15 | 27.11 | 15.2656 | 0.05196 | 2.1319 | 0.1229 | 0.3251 | 1.0235 | 0.3056 | 0.01073 | 8.4906 | 0.02094 |
| 16 | 28.91 | 16.3051 | 0.05550 | 2.2767 | 0.1279 | 0.3462 | 1.0930 | 0.3264 | 0.01146 | 9.0687 | 0.02236 |
| 17 | 30.72 | 17.3476 | 0.05904 | 2.4218 | 0.1329 | 0.3674 | 1.1628 | 0.3473 | 0.01219 | 9.6485 | 0.02379 |
| 18 | 32.52 | 18.3926 | 0.06260 | 2.5674 | 0.1380 | 0.3886 | 1.2327 | 0.3682 | 0.01293 | 10.2298 | 0.02522 |
| 19 | 34.33 | 19.4404 | 0.06617 | 2.7133 | 0.1430 | 0.4099 | 1.3028 | 0.3892 | 0.01366 | 10.8126 | 0.02666 |
| 20 | 36.13 | 20.4911 | 0.06974 | 2.8596 | 0.1481 | 0.4312 | 1.3731 | 0.4103 | 0.01440 | 11.3970 | 0.02810 |
| 21 | 37.94 | 21.5444 | 0.07333 | 3.0063 | 0.1532 | 0.4526 | 1.4436 | 0.4313 | 0.01514 | 11.9828 | 0.02955 |
| 22 | 39.74 | 22.6007 | 0.07692 | 3.1534 | 0.1583 | 0.4740 | 1.5143 | 0.4525 | 0.01588 | 12.5703 | 0.03100 |
| 23 | 41.55 | 23.6597 | 0.08053 | 3.3008 | 0.1634 | 0.4955 | 1.5852 | 0.4737 | 0.01663 | 13.1593 | 0.03245 |

Reproduced from Barnes, H. *J. exp. Biol.* 31, 582-8, 1954. Refer to this paper for further tables and other data on the ionic composition of sea-water.

## Culture Solutions for Flowering Plants

*Pfeffer*

| | |
|---|---|
| $Ca(NO_3)_2 . 4H_2O$ | 4g |
| $KNO_3$ | 1 " |
| $MgSO_4 . 7H_2O$ | 1 " |
| $KH_2PO_4$ | 1 " |
| KCl | 5 " |
| $FeCl_3$ | Trace |
| Distilled water. | 3-71. |

In Knop's solution, the KCl is omitted, and the other substances are dissolved in 12 litres of water. For solutions deficient in one element replace : Ca by K, K by Ca, Mg by K, $NO_3$ by Cl, $SO_4$ by Cl, $PO_4$ by $SO_4$; omit $FeCl_3$.

## Artificial Sea-Water

Calculated from Barnes' data.
Chlorinity = 19%
Salinity = 34.33%.

| | g | |
|---|---|---|
| NaCl | 23.991 | |
| KCl | 0.742 | |
| $CaCl_2$ | 1.135 | (CaCl2. $6H_2O$ 2.240) |
| $MgCl_2$ | 5.102 | (MgCl2. $6H_2O$ 10.893) |
| $Na_2SO_4$ | 4.012 | (Na2$SO_4$.10$H_2O$ 9.100) |
| $NaHCO_3$ | 0.197 | |
| NaBr | 0.085 | (NaBr. $2H_2O$ 0.115) |
| $SrCl_2$ | 0.011 | |
| $H_3BO_3$ | 0.027 | (SrCl2. $6H_2O$ 0.018) |

Dissolve in distilled water and make up to 1 litre.

## Vapour Pressure and Osmotic Equivalence of Sea-Water at 25°C

| % Cl | *R* | NaCl | KCl | CaCl2 | MgCl2 | MgSO4 | $Na_2SO_4$ | Sucrose | Urea | V.P. lowering | Osmotic pressure (atm.) |
|---|---|---|---|---|---|---|---|---|---|---|---|
| 10 | 0.02861 | 0.2861 | 0.2908 | 0.2039 | 0.2005 | 0.5056 | 0.2374 | 0.5065 | 0.5400 | 0.00946 | 12.87 |
| 11 | 0.02869 | 0.3156 | 0.3211 | 0.2240 | 0.2199 | 0.5597 | 0.2643 | 0.5560 | 0.5965 | 0.01042 | 14.19 |
| 12 | 0.02877 | 0.3452 | 0.3516 | 0.2441 | 0.2393 | 0.6138 | 0.2918 | 0.6053 | 0.6534 | 0.01139 | 15.51 |
| 13 | 0.02885 | 0.3751 | 0.3825 | 0.2642 | 0.2588 | 0.6675 | 0.3196 | 0.6546 | 0.7112 | 0.01237 | 16.85 |
| 14 | 0.02893 | 0.4050 | 0.4134 | 0.2841 | 0.2780 | 0.7206 | 0.3477 | 0.7040 | 0.7695 | 0.01334 | 18.19 |
| 15 | 0.02901 | 0.4352 | 0.4447 | 0.3043 | 0.2975 | 0.7738 | 0.3762 | 0.7534 | 0.8285 | 0.01433 | 19.55 |
| 16 | 0.02908 | 0.4653 | 0.4760 | 0.3243 | 0.3165 | 0.8264 | 0.4051 | 0.8025 | 0.8880 | 0.01532 | 20.91 |
| 17 | 0.02916 | 0.4957 | 0.5077 | 0.3445 | 0.3356 | 0.8786 | 0.4347 | 0.8516 | 0.9482 | 0.01631 | 22.28 |
| 18 | 0.02924 | 0.5263 | 0.5397 | 0.3645 | 0.3546 | 0.9300 | 0.4648 | 0.9008 | 1.010 | 0.01732 | 23.66 |
| 19 | 0.02932 | 0.5571 | 0.5719 | 0.3845 | 0.3738 | 0.9803 | 0.4954 | 0.9497 | 1.071 | 0.01833 | 25.06 |
| 20 | 0.02940 | 0.5880 | 0.6043 | 0.4044 | 0.3929 | 1.028 | 0.5264 | 0.9982 | 1.133 | 0.01936 | 26.47 |
| 21 | 0.02948 | 0.6191 | 0.6370 | 0.4243 | 0.4122 | 1.076 | 0.5578 | 1.047 | 1.197 | 0.02039 | 27.89 |
| 22 | 0.02956 | 0.6503 | 0.6698 | 0.4440 | 0.4313 | 1.123 | 0.5896 | 1.095 | 1.260 | 0.02142 | 29.33 |

The column headed v.p. lowering gives the relative pressure lowering $\Delta p/p^o = (p^o - p)/p^o$, where $p$ is the vapour pressure of the sea-water and $p^o$ is the vapour pressure of pure water. $p^o = 23.756$ mm at 25°C.

From Robinson, A.R. : *J. mar. biol. Ass.* U.K. 33, 449. 1954.

## Solubility of Nitrogen, Oxygen, Carbondioxide and Air in Water

The table gives the volume of gas, in ml dissolved in 1,000 ml of water at a total pressure (i.e., partial pressure of gas plus aqueous vapour pressure at the temperature stated) of 760 mm of mercury. (It is assumed that one gram molecule of the gas occupies 22,411.5 ml at S.T.P.)

| Temp. °C | Nitrogen | Oxygen | Carbon dioxide | Air* | Temp. °C | Nitrogen | Oxygen | Carbon dioxide | Air* |
|---|---|---|---|---|---|---|---|---|---|
| 0 | 23.53 | 48.63 | 1.704 | 29.18 | 16 | 17.30 | 34.83 | 1,025 | 20.14 |
| 1 | 23.03 | 47.48 | 1.642 | 28.42 | 17 | 17.04 | 34.23 | 997 | 19.75 |
| 2 | 22.54 | 46.38 | 1.586 | 27.69 | 18 | 16.77 | 33.65 | 970 | 19.38 |
| 3 | 22.08 | 45.32 | 1.533 | 26.99 | 19 | 16.53 | 33.10 | 945 | 19.02 |
| 4 | 21.61 | 44.28 | 1.483 | 26.32 | 20 | 16.29 | 32.56 | 921 | 18.68 |
| 5 | 21.18 | 43.30 | 1.438 | 25.68 | 21 | 16.07 | 32.01 | 898 | 18.34 |
| 6 | 20.74 | 42.36 | 1.395 | 25.06 | 22 | 15.85 | 31.48 | 873 | 18.01 |
| 7 | 20.32 | 41.46 | 1.352 | 24.47 | 23 | 15.65 | 30.96 | 848 | 17.69 |
| 8 | 19.91 | 40.60 | 1.306 | 23.90 | 24 | 15.45 | 30.45 | 825 | 17.38 |
| 9 | 19.53 | 39.77 | 1.264 | 23.36 | 25 | 15.24 | 29.96 | 803 | 17.08 |
| 10 | 19.17 | 38.96 | 1.223 | 22.84 | 26 | 15.05 | 29.49 | 782 | 16.79 |
| 11 | 18.82 | 38.21 | 1.186 | 22.34 | 27 | 14.87 | 29.04 | 762 | 16.50 |
| 12 | 18.49 | 37.47 | 1.150 | 21.87 | 28 | 14.69 | 28.50 | 742 | 16.21 |
| 13 | 18.17 | 36.77 | 1.119 | 21.41 | 29 | 14.51 | 28.17 | 725 | 15.92 |
| 14 | 17.86 | 36.10 | 1.087 | 20.97 | 30 | 14.36 | 27.77 | 707 | 15.64 |
| 15 | 17.58 | 35.45 | 1.057 | 20.55 | | | | | |

* The per cent oxygen in dissolved air at $t$°C (from 0-30°C) is given by the expression 34.91 - 0.044$t$.

## Injection Masses

*Carmine* (Carleton). Mix 5 g carmine with 10 ml distilled water, and leave for some hours (overnight). Add 100 ml strong ammonia and shake to dissolve the carmine. Filter twice through glass wool. Keep at 37°C until all smell of ammonia has completely disappeared (may take several days). Filter through glass wool.

*Prussian blue* (Ranvier). Pour a concentrated feric sulphate solution *gradually* into a concentrated solution of potassium ferrocyanide; this produces a precipitate of prussian blue. Place a funnel and filter paper underneath a felt strainer. Filter and wash the precipitate for some days on the felt strainer. When the liquid passing through the second filter becomes blue, the now soluble precipitate is dissolved in distilled water.

For 'warm' injection mases, add 25 g gelatine and a crystal of thymol to the carmine solution. Mix the prussian blue with 4% gelatine at 37°C instead of distilled water. Kill the animal with chloroform and amyl nitrite. Add a little amyl nitrite to the normal saline used in washing out the blood vessels, and to the injection mass.

*Indian Ink*

*Rubber latex*

## Fixatives

The number of fixatives is legion. The recipes for some commonly used ones are given here.

| | | | |
|---|---|---|---|
| *Acetic Alcohol* | | Bouin (alcoholic) | |
| Glacial acetic acid | 1 part | Picric acid | 1 g |
| Absolute alcohol | 3 parts | Formaldehyde, 40% | 60 ml |
| | | Glacial acetic acid | 15 ml |
| | | Alcohol (80%) | 150 ml |
| *Bouin* | | Carnoy | |
| Picric acid, sat. aq. sol. | 75 ml | Glacial acetic acid | 1 part |
| Formaldehyde, 40% | 25 ml | Absolute alcohol | 6 parts |
| Glacial acetic acid | 5 ml | Chloroform | 3 parts |

*Champy*

| | |
|---|---|
| Potassium dichromate, 3% aq. sol | 7 ml |
| Chromic acid, 1% aq. sol. | 7 ml |
| Osmium tetroxide, 2% aq. sol. | 4 ml |

*Chromo-nitric*

| | |
|---|---|
| Nitric acid, 10% | 4 parts |
| Absolute alcohol | 3 parts |
| Chromic acid, 0.5% aq. sol. | 3 parts |

*Flemming (strong)*

| | |
|---|---|
| Chromic acid, 1% aq.sol. | 15 parts |
| Osmium tetroxide, 2% aq. sol. | 4 parts |
| Glacial acetic acid (add just before use) | 0-1 part |

*Flemming (weak)*

| | |
|---|---|
| Chromic acid, 1% aq. sol. | 125 ml |
| Osmium tetroxide, 2% aq. sol. | 100 ml |
| Glacial acetic acid | 1 ml |
| Distilled water | 775 ml |

*Formol-alcohol*

| | |
|---|---|
| Alcohol, 95% | 15 parts |
| Formaldehyde, 40% | 6 parts |
| Glacial acetic acid | 1 part |
| Distilled water | 30 parts |

*Susa*

| | |
|---|---|
| Mercuric chloride | 45 g |
| Trichloracetic acid | 20 g |
| Glacial acetic acid | 40 ml |
| Fomaldehyde (40%) | 200 ml |
| Sodium chloride | 5 g |
| Distilled water | 800 ml |

Formol-calcium

| | |
|---|---|
| Formaldehyde, 40%, | 10 ml |
| Calcium chloride (anhydr.) 10% aq. sol. | 10 ml |
| Distilled water | 80 ml |
| $CaCO_3$, powdered, slight excess | |

Gilson

| | |
|---|---|
| Nitric acid, conc. | 15 ml |
| Glacial acetic acid | 4 ml |
| Mercuric chloride | 20 g |
| Alcohol 60% | 100 ml |
| Distilled water | 880 ml |

Helly

| | |
|---|---|
| Potassium dichromate, 3% aq. sol. | 50 ml |
| Mercuric chloride, sat. aq. sol. | 50 ml |
| Sodium sulphate | 1 g |
| Formaldehyde, 40% (Add just before use) | 5 ml |

Schaudinn

| | |
|---|---|
| Mercuric chloride, sat. aq. sol. | 100 ml |
| Absolute alcohol | 50 ml |
| Glacial acetic acid | 5 drops |

Smith

| | |
|---|---|
| Potassium dichromate | 0-5 g |
| Glacial acetic acid | 2-5 ml |
| Distilled water | 100 ml |
| Formaldehyde, 40% (Add just before use) | 10 ml |

Zenker

| | |
|---|---|
| Mercuric chloride | 5-0 g |
| Potassium dichromate | 2.5 g |
| Sodium sulphate (10 $H_2O$) | 1.0 g |
| Distilled water | 100 ml |
| Glacial acetic acid (add just before use) | 5 ml |

## Macerating Fluids

30 *Per cent Alcohol* (Ranvier) 24 hours.

*Boric acid* (Goodrich) Saturated solution in saline, (sea- water or Ca-free sea water for marine forms) plus two drops of Lugol's iodine per 25 ml of boric acid solution. 2-3 days. (Lugol's iodine : 1 per cent sol. of iodine in 2 per cent aqueous potassium iodide).

*Chromic acid* 0-2 per cent (Ranvier) 24 hours. For animal tissues.

*Chromic acid* 5 per cent (Priestly) 24 hours. For plant tissues.

*Nitric acid* 20 per cent 24 hours.

*Nitric acid* (10 per cent) plus chromic acid (10 per cent), in equal volumes (Jeffery). 12-24 hours at 35°C. For plant tissues.

To obtain viable cell suspensions of embryonic material see Moscona. *A. Exp. Cell Res.* 3, 535-9. 1952. Zwilling, E. *Science.* 120. 219. 1954.

## Decalcifying Fluids

1-5 per cent Nitric acid in 5 per cent formol.

Sulphurous acid (saturated solution of $SO_2$).

*Susa's fixative.*

10 per cent Versene, or other similiar chelating agent (Birge and Imhoff).

### SOFTENING HARD MATERIALS

*Lendrum's method.* Wash out fixative, and soak for 1-3 days in 4 per cent aqueous phenol. Then dehydrate as usual.

Material may be softened in a wax block. Expose tissue and soak in the following fluid (Baker).

| | |
|---|---|
| Water | 36 ml |
| 96 percent alcohol | 54 ml |
| Glycerol | 10 ml |

For plant material, replace the glycerol with glacial acetic acid (Gifford).

## Dehydrating Agents

Ethanol, Acetone, Iso-propanol, *n*-propanol.

## Clearing Agents

| Substances | Lowest grade of alcohol from which objects may be transferred* | Remarks |
|---|---|---|
| Amyl acetate | 95% | No hardening of tissues |
| Aniline oil | 70% | Not after osmic fixation. Does not mix well with paraffin |
| Beechwood creosote | 90% (?) | Good for whole mounts |
| Benzene | 98% | Rapidly penetrates tissues |
| Cedarwood oil | 95% | Mixes slowly with paraffin |
| Chloroform | 95% | Penetrates tissues slowly |
| Clove oil | 75% | Makes tissues britle. Removes some strains |
| Methyl benzoate | 98% (?) | |
| Terpineol | 80% | Good for hard material |
| Toluol | 98% | Rapidly penetrates tissues |
| Xylol | 98% | Rapidly penetrates tissues, but makes them brittle |

*Assuming clearing agent to be anhydrous.

## Combined Dehydrating and Clearing Agents

*Dioxane* (di-ethylene dioxide). Stand over calcium oxide. A 1 : 1 mixture with ethanol dissolves celloidin.

*Cellosolve* (ethylene glycol mono-ethyl ether). Dissolves celloidin.

## Mounting Media and their refractive indices

MISCIBLE WITH WATER

| *Temporary* | | Permanent | |
|---|---|---|---|
| Distilled water | 1.336 | Glycerine jelly | 1.45-1.47 |
| Sea-Water | 1.343 | Farrant's medium | |
| Glycerine 50% | 1.397 | Faure's gum chloral | |
| Glycerine | 1.473 | Apathy's medium | 1.52 |
| Diethylene glycol | 1.448 | | |
| Sat. sol $HgI_2$ and KI in water | 1.68 | | |

MISCIBLE WITH ALCOHOL

| | |
|---|---|
| Euparal | 1.482 |
| Alcohol balsam | |
| Dioxane balsam | |

MISCIBLE WITH CLEARING AGENTS

| *Temporary* | | Permanent | |
|---|---|---|---|
| Aniline oil | 1.580 | Balsam (Solid) | 1.535 |
| Carbon disulphide | 1.630 | Balsam (xylol) | 1.524 |
| Castor oil | 1.490 | D.P.X. | 1.51 |
| Cedarwood oil | 1.51-1.52 | Gum Dammer | 1.520 |
| Clove oil | 1.533 | Gurr's neutral mounting medium | 1.51 |
| Creosote | 1.538 | | |
| Methyl benzoate | 1.517 | Xam | 1.491 |
| Olive oil | 1.473 | | |
| Paraffin (liquid) | 1.471 | | |
| Xylol | 1.497 | | |
| Naphthalene monobromide | 1.660 | | |
| Solution of sulphur in carbon disulphide | 1.750 | | |
| Hyrax | 1.822 | | |

# BIBLIOGRAPHY


Anonymous (1987). Approved methods of the AACC. American Association of Cereal Chemists, St. Paul, MN.

APHA (American Public Health Association) (1975). Standard methods for examination of water and waste water; 14th ed. Prepared and published by APHA. AWWA. 1015 Eighteen street, NW. Washington D.C. 20036 pp. 1193.

Arnon DJ (1949). Copper enzymes in isolated chloroplasts. Plant Physiol. 24 : 1-25.

Aspinall D and Paleg LG (1981). Proline accumulation : Physiological aspects. In : The physiology and Biochemistry of drought resistance in plants (Eds. Lg Paleg and D Aspinall). pp. 205-241. Academic Press, New York.

Baker DL and Shur NH (1982). Atomic absorption and flame emission spectrometry, pp. 13-27. In : (Ed. AL Page), Methods of Soil Analysis. Part 2. Chemical and Microbiological Properties. Agronomy Monograph No. 9. American Society of Agronomy, Madison, WI.

Bates LS, Waldren RP, Teare ID (1973). Rapid determination of proline for water stress studies. Plant Soil 39 : 205-207.

Baylis AD (1990). Economic aspects of plant growth regulators. In : Plant Growth Substance (Eds. Pharis, RP, Rood SB) Springer Verlag. Berlin, pp. 574-584.

Blackman GE (1968). The application of concept of growth analysis to the assessment of productivity. In : Kc Kordt FF (ed) Functioning of Terrestrial Ecosystems at primary production level. pp 243-260, Proceeding of Copenhegan Symposium, UNESCO.

Bock R (1978). Handbook of Decomposition Methods in Analytical Chemistry. International Textbook Company, Glasgow, Scotland.

Bower LA and Gschwend FB (1952). Exchangeable cation analysis of saline and alkali soils. Soil Sci. 73 : 251-261.

Bradstreet RB (1965). The Kjeldahl Method for Organic Nitrogen. Academic Press, New York.

Burris RH (1972). Nitrogen fixation assay method and techniques. Methods Enzymol., 24 : 415-431.

Burton K. (1956). A study of the conditions and mechanism of the diphenylamine reaction for the colorimetric estimation of deoxyribonucleic acid. Biochem. J. 62 : 315-323.

Dubois M, Gilles KA, Hamilton JK, Roberts PA and Smith F (1956). Colorimetric method for determination of sugars and related substances. Anal. Chem., 28 : 350-356.

Fletcher RA and Drexlure DM (1980). Interactions of dichlofop-methyl and 2,4-D in cultivated oats (*Avena sativa*). Weed Sci 28 : 363-366.

Gorsuch TT (1970). The Destruction of Organic matter. Pergamon Press, Oxford, England.

Hanlon EA (1992). Determination of potassium, calcium and magnesium in plants by atomic absorption techniques. pp. 33-36. In : (Ed. Plank CO), Plant Analysis Reference Procedures for the Southern Region of the United States. Southern Cooperative Series Bulletin 368. University of Georgia, Athens.

Ikan R (1969). Natural products : A Laboratory Guide. Academic Press, New York.

Jackson ML (1962). Soil Chemical Analysis. Asia Publishing House. pp. 452-465.

Jackson ML (1973). Soil chemical analysis. Prentice Hall, India (Pvt.) Ltd., New Delhi pp. 1-497.

Kirk JTO and Allen RL (1965). Dependence of chloroplast pigment synthesis on protein synthesis : Effect of actidione. Biochem Biophys Res Commun 21 : 523-530.

Lowry OH, Rosebrough NJ, Farr AL, Randall RJ (1951). Protein measurement with folin phenol reagent. J Biol Chem 193 : 265-275.

McDonald and Chen (1965). Anal Biochem 10 : 175-177.

Mc Donald MS (1978). Asimple and improved method for the determination of microgram quantities of N in plant materials. Ann. Bot. N.S. 42, 363-366.

McLean AJ (1976). Cadmium in different plant species and its availability in soils as influenced by organic matter and additions of lime., P, Cd, and Zn. can. J. Soil Sci. 56 : 129-138.

Mirhatum and Hume, DJ (1976). Relations between nitrate reductase activity and nitrogen accumulation in soybean. Can. J. Plant Sci., 56 : 377-384.

Murashige T and Skoog F (1962). A revised medium for rapid growth and bioassay with tobacco tissue culture, Physiol. Plant, 15 : 473-497.

Pendiras AK and Pendiras H (1984). Trace elements in soils and plants. CRC Press Inc. Boca, Raton, Florida.

Raina SK (1989). Tissue culture in rice improvement, status and potential. Advances in Agronomy, 42 : 339-398.

Schaffert RR and Kingsley GR (1955). A simple method for the determination of reduced dihydro total ascorbic acid in biological materials. J Biol Chem 212 : 59-68.

Shen J, Li M, Chen Y and Zhang Z (1983). In : Cell and tissue culture techniques for cereal crop improvement. Science Press, Beijing, 183-205.

Srivastava RC and Mathur SN (1980). Nodulation and NR activity in nodules and acid leaves of black gram (*V. mungo*) as affected by varying day length. Indian J. Exp. Biol 18 : 300-302.

Strain HH and Svec WA (1966). Extraction, separation, estimation and isolation of chlorophyll. In : LP Vernon, GR Seely (eds). The Chlorophylls. Academic Press, New York pp. 21-66.

Tanaka K and Sugahara K (1980). Role of superoxide dismutase in defense against $SO_2$ toxicity and an increase in superoxide dismutase with $SO_2$ fumigation. Plant Cell Physiol. 21 : 601-611.

Vessey JK, and Waterer J (1992). In search of the mechanism of nitrate inhibition of nitrogenase activity in legume nodules : Recent developments. Physiol. Plant., 84 : 171-176.